The Guardian of Technical Excellence

THE GUARDIAN OF TECHNICAL EXCELLENCE

Jeff Deckelbaum

ACKNOWLEDGEMENTS. This book is written around my work experiences, and files of personal notes and artifacts collected through the years. I want to sincerely thank the following people from Boeing for the time they took to read, edit and provide me with their inputs and recommendations to ensure complete accuracy throughout the manuscript; Charles Dabundo, Patrick Dolan, Philip Dunford, Chadd Fleming, David Koopersmith and Lori Middendorf. In addition, I want to thank other Boeing subject matter experts who contributed their inputs on selected topics of interest.

Also I want to thank my mother, Sandra Deckelbaum, who helped me by providing and verifying information about the early years of my grandparents' lives, and with editing.

A very special thanks to my friend and coworker, Mark Bass who worked with me on this project from the start and provided me with invaluable input and advice during the process of developing and reviewing the manuscript for content and accuracy.

In addition, I want to thank Brayton Harris. Brayton is a retired Naval Officer, a former aerospace executive and a distinguished author in his own right who helped me improve the readability and flow of this book, and with the publishing process. Brayton also guided and mentored me through the rigorous process of what it takes to author a book.

DEDICATION

To my parents, Sandra and Stanley, who raised and encouraged me to pursue my lifelong dreams;

To my wife Kathy, the love of my life who helped me achieve those dreams;

To our children, Jason, Julie and Dan; I am so very proud of the adults they have grown up to be, and,

To our grandchildren, Annie, Jack, Caroline and Corbin who I love spending time with, and who will carry on our family legacy in the future.

TABLE OF CONTENTS

PREFACE

Are you an early or mid-career engineer working in industry today thinking about how can you learn from the experiences of others? Perhaps you might want to strengthen and broaden your technical skills to advance your career, or gain insight on how to best transition into a position of technical leadership, or become a program or business unit level Chief Engineer. Maybe you have heard the mantras "First Time Quality" and "Technical Excellence" and think, "What does that really mean or how does that apply to me?"

If you would like answers to these questions and many more, then this book might have something to offer you.

- What is "working in industry" like?

- Am I prepared to transition from student to professional?

- What will my boss and teammates expect of me?

- How can I gain clarity on specific engineering concepts and why they are important?

- When is the best time to get an advanced degree?

- What are the benefits of a business degree?

- As a business leader, how can I learn what other industry leaders are expecting from engineers?

- As a high school student considering an engineering

career, how can I learn more about what engineers do and what is expected?

The Guardian of Technical Excellence addresses the complexity of designing, building, testing and fielding highly capable military aircraft. Best practices and lessons learned are shared around the unique challenges engineers in industry face, the decisions that engineers make and the types of risks that we manage. There are significant implications if we don't get the engineering right. This includes customers not having capabilities when they need them, to accomplish their missions, and facing cost overruns that can adversely affect the government, prime contractors and suppliers.

As an example, this book addresses solving an issue cited by a U.S. Air Force procurement official, "Service development programs in the Pentagon are—on average—taking about 40 percent longer than the planned five years expected to mature into the production stage, largely due to problems with software."[1]

The goals of this book—and what readers can expect to learn—are:

1. A differentiated set of best practices and lessons learned from the perspective of a military aircraft Chief Engineer—combining the need for a strong technical foundation with a commitment to technical excellence. Our focus is on military aircraft weapon systems, but the lessons are applicable to many other industries.

2. What is "Technical Excellence"? How can it be operationalized through an understanding of the

Aircraft Product Life Cycle process, backed up with relevant Case Studies.

3. I provide answers to typical questions I have been asked while engaging with teams, teaching classes at Boeing's Leadership Center, or meeting with a group of eager and ambitious emerging leaders:

> a. How did you get to where you are in the company?

> b. What were your biggest work challenges? How did you handle them? What was the outcome?

> c. Can you please share your thoughts on how to achieve work-life balance?

4. The importance of aligning and integrating people and teams around common goals and objectives.

5. Creating a culture where: it's okay to bring bad news forward; to not shoot the messenger; and where asking for help is encouraged.

6. The importance of demonstrating not only performance, but also core values, such as ethical decision-making, trust, respect and integrity.

7. Encouragement for students to consider engineering as a career, especially those with an interest in mathematics and science.

My credentials include more than 37 years of work experience, predominately in Engineering, and also Program Management, Operations, Quality, and Information Systems. I led Engineering for the two business units within The Boeing Company's Defense, Space & Security business for nearly the last 8 years of my career. These business units develop, build and support Boeing's military aircraft for the U.S. and allies. Up to this writing, I was privileged to be the only person to lead both Boeing Military Aircraft (Aircraft Design/Production) and Global Services & Support (Post-delivery support, which included Design/Modification) Engineering organizations.

My education consists of Bachelor and Master of Science degrees in Aeronautical, Aerospace and Astronautical Engineering from Purdue University (1977) and San Diego State (1980), respectively. I earned a Master of Business Administration from Washington University (1994). I attended Notre Dame's Integral Leadership Program in 2004. ∎

---

1. Butler, Amy, *Defense Development Timelines Thwarted by Poor Software Planning*, p.4 *Aviation Week's* Intelligence Network's Aerospace Daily and Defense Report, 10 July 2015.

INTRODUCTION

I am an engineer. Engineers, through their knowledge, imagination, innovation, creativity, and approach to problem solving, together with emerging technologies, are providing customers with products that make the world go around. Recent examples of product innovation include the smart phone that many can't live without; the latest fitness band to measure how many steps walked in a day; Tesla's high performance electric car; advances in medical scanning equipment that help people live longer through early detection of illness or disease; and the latest generation of aircraft connecting and protecting America and its allies. Other advances include rapid prototyping, which can bring capabilities to market faster than ever before, immersive development technology, which helps customers understand and shape new products, and 3D printing that provides realistic replicas.

Engineers tend to be, well, compulsive at times, focus on minutiae, driven by facts and data . . . traits that may drive coworkers, friends and family to distraction, maybe even frustration. But, in defense of engineers . . . if we don't pay attention to the details and we don't get it "right," the bridge falls down, the ship capsizes, or the airplane won't fly.

As a military aircraft chief engineer, I acquired the knowledge to design, build, test, and field incredibly complex weapon systems—and I know how to assemble, lead and manage the teams to make it all work. This includes the critical processes and tools required to help engineers do their jobs. Not just the folks at the front end, the "in the beginning" bit, focused on aircraft configuration, aerodynamic loads, materials, weights, and performance, but also the folks who do the detailed design, double-check computations and certify, "ready to fly." As a chief engineer, my responsibility doesn't end until the "system"—often, a full-up high-performance military airplane—has been built, delivered and in service for many, many years.

Aircraft programs are generally organized into Integrated Product Teams (IPTs), which include engineers and other skilled people to translate customer requirements into a product through disciplined processes of design, test and support after delivery. These teams have full responsibility, authority and accountability (RAA) for their product, such as, a wing or a fuselage. This IPT concept encourages teamwork, real time decision-making and enables adapting to whatever change or adjustments in the plan are needed as product development evolves.

However, all too often, these IPTs can end up as a collection of stovepipes, of teams each focused on one aspect or another. Requirements may not all be clear, or information at the start of design product such as aerodynamic loads applied to the wing is in a preliminary stage. Early in the process, interfaces are not always defined to the level of detail or maturity required. The design comes together at the top, but sometimes with not enough interaction and coordination at the bottom, the beginning or middle, of the project. If and when this happens, engineering rework happens, resulting in cost overruns and schedule delays.

And that's where I have something to offer, how to assemble, align and integrate "teams," not through some amorphous happy-feely thing, but as integral and vital parts of a program. Yes, I know, the world of business is full of experts on team-building, some of whom offer—for example—off-site weekend "exercises" where four or five employees team up for a Treasure Hunt Adventure, which begins as a competition but morphs into a cooperative endeavor as all teams must work together to find the key. All had a great time because, well, because of the accomplishment and the "bonding" with each other.

Wonderful. That makes for grand celebration at the local pub, at least until the thrill wears off. But . . . how do you measure this as "success," in a program sense of the word?

I've spent a lifetime in this field, and I'd like to bring the focus of an engineering executive—yes, someone who knows when to be obsessed at times with detail but not blind to everything else that goes into ensuring success for the programs and the business. I call it, the pairing of leadership with technical excellence.

You get the idea. Read on . . . ■

NOTE: Because some "engineering speak" may seem, at first, a bit puzzling, and, because there are so many acronyms (OPEVAL, V/STOL) and initialisms (GAO, DFMA, LRIP, ICD, IPT) used in the business . . . you get the idea, I have added a glossary, easily found in a last few pages of this book.

CHAPTER ONE
SETTING THE CONTEXT

*For me context is the key—from that comes
the understanding of everything.*

Kenneth Noland
Artist

To provide context for the chapters ahead, it's important to understand the business of aerospace and the role of a major aircraft engineering and manufacturing company. The Boeing Company has two primary profit and loss businesses, Defense, Space and Security (BDS), and Commercial Airplanes. Nearly my entire career was spent in two business units within BDS, those being Boeing Military Aircraft (BMA) and Global Services & Support (GS&S).[1] BMA develops, builds, tests and delivers new aircraft weapon systems. GS&S provides innovative logistics, sustainment, and after-delivery support solutions. Within these two business units are Divisions, comprised of a portfolio of programs, that align with our domestic and international military customers

Military programs have enormous complexities and it takes leadership and teams of skilled people to acquire, develop, build and support these aircraft weapon systems. The Acquisition phase of programs can take years, depending on customer needs and priorities, as well as congressional approval and funding.

Development of larger programs can take upwards of ten years, and programs can be in production for as many years as customers want to buy the aircraft in economic quantities (F-15 fighter aircraft have been in production for more than 40 years.) The venerable B-52 bomber and KC-135 aerial refueling tanker have been supported since the 1950s, and continue to be part of today's force structure providing tremendous capability to the warfighter.

People have come and gone through multiple generations, and skill mix requirements have changed as well, as programs have entered and exited the phases of development, build and support. Organizational constructs have changed and evolved with the intention of increasing efficiency and effectiveness, while driving decision-making down to the lowest or team level.

Programs, however, are not the only entities held accountable for delivering products and services to customers that meet their requirements and specifications, within cost and schedule commitments.

Within BDS there are functional organizations, such as Engineering, in which leaders are responsible to provide people, processes, and tools, and ensure functional or technical excellence to each and every program through a matrix organization, described in further detail in Chapter Seven. Technical Excellence is a key theme throughout the book.

Technical Excellence is a journey that combines leadership, culture, strategy, organizational constructs and process, together with engineering domain knowledge and expertise, to support military aircraft programs, and protect the company from anything that conflicts with core values and principles.

As noted, above, my goals, here, are to share best practices and lessons learned through my education and work experience. Some major themes by chapter are as follows:

- Chapter One: Setting the Context

  o Sets context for the reader through some basic information on The Boeing Company's Defense business and the business of aerospace.

  o Defines "Technical Excellence."

- Chapter Two: Fulfilling my Passion

  o I share my personal feelings on my last workday as a senior engineering executive.

  o And, describes what life was like for me toward the end of my career, to include a balance of highlights and challenges faced.

- Chapter Three: How and Why I Became an Engineer

  o Provides background on my family roots and growing up years, beginning when I was 10 years old.

  o Then, explains how I became interested in airplanes, and combined with strengths in mathematics and science, how I followed my passion to become an aeronautical engineer.

- Chapter Four: Building a Strong Technical Foundation

  o Emphasizes why building a strong technical

foundation is a prerequisite to advancement into a leadership position—and that it doesn't happen overnight.

o Addresses when is the best time to pursue a Masters Degree in Engineering and in Business Administration and why.

- Chapter Five: Early to Mid-Career Transition

o Highlights best practices and lessons learned from my own work experiences as I transitioned from early to mid-career, which includes working with people, teams and customers.

o And, why it's important to understand the guts of the products you are responsible for and the processes used to develop, build, operate and sustain them after delivery.

- Chapter Six: People

o Defines the talent needed to execute complex engineering programs and characteristics required for success.

o How to select, motivate, inspire, develop and evaluate talent.

o Discusses 'diversity', and how diversity in teams can be a competitive advantage, and characteristics of the work environment to maximize diversity of thought.

- Chapter Seven: Teams and Team Structures

  o How to align people in organizational or team structures to maximize efficiency and effectiveness.

  o The use of Technical Lead Engineers to "advise, assist and check work" as a means of ensuring first time quality.

- Chapter Eight: Aircraft Program Development and Demonstration

  o Provides a step-by-step executive summary of the Product Life Cycle and Systems Engineering processes used to guide program Development and Demonstration.

  o What is Design for Manufacturing and Assembly (DFMA), Ergonomics and Safety, and Supportability and why are they such important considerations during the product development process for those that manufacture (build team) and support (warfighters) aircraft weapon systems.

  o Provide the warfighters with the capabilities they need, when they need it, and on budget or cost.

- Chapter Nine: Managing Complexity

  o Defines Sources of Complexity and offers actions on how to simplify it and manage it.

  o Provides descriptions on key processes required to execute a statement of work, such as Lean+, Root Cause and Corrective Action, and Earned Value Management.

- Chapter Ten: Selected Case Studies

  o Provides additional insight through five relevant case studies on selected programs where I had a leadership role as a business unit chief engineer. These case studies describe a specific situation, the challenges, actions taken and the outcome.

  o Examples at different phases of the Product Life Cycle are utilized to highlight similarities and differences of best practices and lessons learned.

- Chapter Eleven: Operationalizing Technical Excellence

  o Expands on the theme of Technical Excellence and provides specifics on how to achieve it across organizations.

  o Describes how to establish an Engineering System by leveraging enterprise or corporate-wide resources combined with processes on how to engage teams in a value-added way to flesh out and help mitigate technical risks and issues.

  o Explains the Importance of a Culture of "Help Needed" and "It's okay to bring bad news forward" through all phases of the product life cycle.

  o Highlights that a safe work environment is a more productive work environment and a competitive advantage—and your responsibility to create that safe environment.

- Chapter Twelve: Ethics in Engineering

  o Clearly defines what ethics in engineering is about and the consequence of unethical behavior to you as an individual and to a company.

  o There is NO compromise for product safety, integrity and quality.

- Chapter Thirteen: Career Mapping

  o Provides a framework for career mapping to achieve your professional objectives.

  o Suggests the different types of development opportunities and when to pursue them.

  o Reiterates the need to have a strong technical foundation when advancing into an engineering leadership position of responsibility.

- Chapter Fourteen: Executive Engineering Leadership

  o Describes the responsibility of a Chief Engineer—which includes ensuring both a culture of technical excellence and a work environment built on a foundation of ethics, integrity and quality.

  o Highlights our responsibility to promote science, technology, engineering and mathematics to students, as early as in elementary school. In other words, plant the seeds to ensure a pipeline of engineers for years in the future.

- Chapter Fifteen: Choosing to be Unstoppable

  o Realizing that without your health, you will not be much good for yourself, your family, and your people and teams.

  o The need to have outside interests that you use to decompress from the pressures from work, and life in general.

  o I share my passion for strength and conditioning training, as both my recreational activity of choice, but also to offset the pressures of my job. It's a life-long commitment for me.

- Chapter Sixteen: Call to Action

  o Provides a list of considerations and questions to readers based on the key themes throughout the book.

  o Find two or three things that are most important to you and take action. ∎

---

1. Boeing Military Aircraft and Global Services and Support were the names of the business units at the time of my retirement. The names of organizations changed many times during my career to include before the time when McDonnell Douglas and Boeing became One Company in 1997. My work experiences prior to that time were predominately in Military Fighter Aircraft, synonymous with BMA.

CHAPTER TWO
FULFILLING MY PASSION

Nothing stops the man who desires to achieve.
Every obstacle is simply a course to develop
his achievement muscle. It's a strengthening
of his powers of accomplishment.

Thomas Carlyle
Scottish Philosopher and Teacher

Friday, October 11th, 2013 would be my last day working at The Boeing Company, retiring as Vice President and Chief Engineer for Boeing Military Aircraft (BMA). With accrued and unused vacation, my retirement date was November 23, almost 34 years to the day I started in 1979 at McDonnell Douglas in St. Louis.

Coincidentally, about three weeks prior to my last day at work, I would receive an Outstanding Aerospace Engineer award from my alma mater, Purdue University recognizing my contributions and accomplishments throughout my professional career. Ironic that the timing of this recognition came when it did. When I received the award, I saw some of my professors; they were retired of course, but how cool that was to engage with them on a whole different level some 36 years later.

The last day was filled with emotion. It was the end of a very rewarding career. I would be leaving a company that designed and built the best aircraft in the world, which for years to come will continue to protect and defend our nation and those of our allies. There had been many successes of which I was proud, and a few setbacks along the way from which came some hard-earned lessons. It was tough to say goodbye to so many people I had worked with for so many years, sharing the journey of change and evolution inside the company. And, in the world.

I led Boeing Military Aircraft Engineering for the last three years of my career, an organization built around our core business of developing and producing military aircraft and weapon systems. My entire career had prepared me for this assignment leading a team of approximately 9,000 people, working for the most part at five sites across the U.S. We handled a $16B annual portfolio of large, complex programs, including fighter aircraft and weapons, rotorcraft, unmanned aircraft systems, mobility, tanker and intelligence, reconnaissance and surveillance (ISR) aircraft.

My passion was leading and working together with people, engaging with customers, achieving technical excellence and delivering results. A career highlight, 2012, was spending two days on the aircraft carrier, USS *John C. Stennis,* CVN 74, talking with senior officers, pilots, maintainers and sailors. Standing on the flight deck when aircraft were being launched and trapped on landing, I felt like I was in the movie, *Top Gun.* Only the star of the movie was the F/A-18 *Super Hornet.*

I had the opportunity to feel the exhilaration of both a carrier landing and catapult takeoff aboard a Grumman C-2 *Greyhound,* a Carrier Onboard Delivery (COD) aircraft. I watched night flight operations, standing on "Vulture's Row," an observation area on one of the upper decks of the aircraft carrier's island. The professionalism, efficiency and discipline of every team member on that ship was so impressive, including how they managed aircraft on the carrier deck and moved supplies from a supply ship onto the carrier. It was all motivating and inspiring to watch.

Throughout my career, I traveled a good part of the world experiencing the cultures of Australia, Israel, and many countries in Europe. There was always a motivation, to deliver results in an ever-changing global environment and dealing with technical challenges. Decisions to make today, plans to make for tomorrow.

My daily routine began very early through most of my career. For three to four mornings each week, my day started at the gym followed by a 12-hour workday in whatever city or country my job took me. I would travel usually two or three weeks out of every month engaging with our people, teams, customers and suppliers. The pace, stress, responsibility and personal accountability associated with the job, though, could sometimes be overwhelming.

On the home front, my wife Kathy and I have been married since we graduated college and now with two grown children, Jason and Julie, working in professional jobs. We have four grandchildren, Annie, Jack, Caroline and Corbin. They were born between 2007 and 2012. We dealt with elderly parent issues; most recently with my father in law passing in 2012 and my dad Stanley passing away a few weeks before my retirement. Life at home was getting busier quickly, and at the same time my job was getting more demanding.

Family was, and still is, most important to me. Work Life balance was getting harder to achieve; in fact I wasn't doing a good job at all. I wanted to spend more time at home with the family. I was already missing out on many fun times with the grandchildren. There just wasn't enough time in the day; and on most days I was in town, I came home exhausted and distracted from the physical and mental demands of the job.

I met and worked with a lot of people, and made lifelong friends. I also had the opportunity to develop and work closely with the person who would be my successor. He was well prepared to come in behind me and I felt good about leaving the team in his capable hands. I wanted to retire on my terms, on a positive note and with my health, and I did.■

Chapter Two Takeaways

● Whatever your chosen career path is, work hard and make a difference.

● Take care of yourself—you will not be good to anyone if you are not healthy

● Retire on your terms; don't let anyone make that decision for you.

CHAPTER THREE
HOW AND WHY I BECAME
AN ENGINEER

A dream becomes a goal when action is taken
toward achieving it.

Unknown

In the summer of 1965, the textile company for which my father worked transferred operations from Philadelphia, to the
small town of Albemarle, N.C., about 35 miles outside of
Charlotte. He was a Plant Manager at a company named Jefferies
Southern Processors.

I was 10 years old and the oldest of three kids, with a younger
brother and sister. My grandparents and most of our extended
family lived in Washington D.C. so we were within a three-hour
drive when we were living in Philly. I had an aunt, uncle and
cousin, and some great childhood friends in Philadelphia that we
would really miss. I remember, though, being excited about moving to the "Queen City of the South" in time to start my 5th grade
school year.

My father was my role model. He was a great dad, treated people with kindness and respect, worked his butt off to be a good
provider and had a work ethic second to none. He received his
Bachelor of Science Degree in Chemistry at Philadelphia Textile

Institute. My mother was a stay at home mom, took care of us kids, and instilled a sense of independence in us, maybe too much when I was growing up.

All of my grandparents came to the U.S. in the early 1900s, escaping from towns in then Russia and Poland, and what is now Ukraine. In his memoirs, my paternal grandfather, Isadore, would write about the horrific brutality of Russian soldiers during the Pogroms around the time of World War I, and about avoiding capture and never having enough food to eat.[1] Together with his mother, five brothers, sister, and my future grandmother, they would come to the U.S. in 1920. He and three of his brothers would own small corner grocery stores in Washington, D.C.

My maternal grandparents were both young children when they and their parents escaped the Russian oppression My mother's father, Joel Novick, would attend medical school at George Washington University and become an Otolaryngologist. All four grandparents would eventually settle in Washington D.C. where my parents were born, raised and then married in December of 1953. My brother and I were also born there.

The Charlotte of 1965 was not a large city, perhaps 250,000 people within the greater Charlotte-Mecklenburg county area. But it was large enough to support a commercial airport, and I soon enough noticed the flights of Eastern Airlines, Delta and Piedmont Airlines. I was fascinated every time I heard a noise of jet aircraft above. The two predominate jet aircraft of the era that flew in and out of Charlotte were the McDonnell Douglas DC-9 and the Boeing 727. Little did I know at the time, I would work for both of those two great companies.

On weekends and holidays, my mother would sometimes drop

my brother and me off at the airport for a few hours. (Did I mention she was thrilled to get rid of us for a few hours?) We would go onto an observation deck and watch the airplanes take-off and land. Sometimes, we could go up into the control tower and talk with the controllers. Can you believe how things have changed?

In the summer between my 9th and 10th grade year, my parents sent all three of us kids to a summer camp in North Carolina, a town on the Neuse River called Arapahoe and not far from the Atlantic Ocean. When settling in on the first day, well I recall hearing a noise from a jet aircraft, much louder than I had ever heard before. I ran outside, and while I didn't know what it was, I was looking at an F-4 Phantom landing directly across the river at the Cherry Point Marine Corps Air Station. For the next five weeks, I watched an awesome display of fighter and other aircraft flying by. At camp in the following two summers, I would watch the military aircraft, every day.

During my junior high years, I started to play tennis and would develop a passion for the game that lasted well into my mid-fifties. I played on a good high school tennis team. In fact, one of my teammates, John Sadri, would go on to play college tennis at North Carolina State University in Raleigh, N.C. He would lose in the finals of the NCAA Division I tournament to John McEnroe in 1978 in what was an unbelievable match. Sadri would later turn pro and rise into the top twenty players in the early 1980s.

When it was time to think about colleges and careers, I knew it was going to be something with airplanes. With strengths and interest in mathematics and science, my dad encouraged me to pursue engineering as a career. We had family friends that graduated from Purdue University, which had an excellent program in Aeronautical and Astronautical Engineering. With Neil

Armstrong and Gus Grissom as Purdue University graduates, I knew what and where I was going to study.

I applied for early admission and was accepted in October of 1972. I have vivid recollection of the day I received the good news. My brother and I each had jobs after school. While driving to work on the day I was accepted (and I didn't know it at the time) we were broadsided at a busy intersection. An elderly person ran through a stop sign and hit the car and spun us around. We joke now that elderly person was probably not so elderly (interesting how your point of view changes over time). Fortunately, no one was seriously injured and my entry to Purdue was not affected.

Like most engineering students, I completed most of my basic mathematics, science and general education courses during my freshman and sophomore years. As I started my junior year, as I got more enthusiastic about my education and what I was learning. I gravitated to structural engineering. I took every undergraduate and graduate course I could fit in, to include composite materials. I also studied among other subjects: aerodynamics, fluid mechanics, air breathing propulsion, stability and control. I also took classes in thermodynamics (from the school of Mechanical Engineering) and electrical circuit theory (from the school of Electrical Engineering.) Call this, "Step One" on the road to building my aerospace technical foundation.

While in college, I learned to live independently, the importance of teamwork, meeting deadlines, and the vital role of personal accountability. I built relationships with students and faculty, all part of my positive experience at Purdue University. Looking back, I wish I might have had the opportunity to spend a summer working as an intern in industry. In those days, internships were few and far between; to the benefit of us all, times have changed.■

Chapter Three Takeaways

• Follow your passion, take it forward into a career. You will spend too much time working to not love what you do.

• If you want to be an engineer, you will need strength and interest in mathematics and science, but also skill in reading, writing, and communication. . . . and you must be ready to remain current with all facets of the global environment in which engineering is practiced.

• High school is where you start to build your technical foundation with mathemaics and science courses and come to understand the vital role of personal accountability.

• Be alert to opportunities for industry internships . . . where you can integrate with what you are learning in school with what is being done in industry.

---

1. Deckelbaum, Isadore, Isadore Deckelbaum Memoirs (Part 1 and Part 2), 1976—1977: http://genealogy.caroldeckelbaum.com/memoir-izzy1.html ; http://genealogy.caroldeckelbaum.com/memoir-izzy2.html (Retrieved 16 Nov, 2015)

CHAPTER FOUR
BUILDING A STRONG
TECHNICAL FOUNDATION

Continuous effort—not strength or intelligence—is the key to unlocking our potential.

Winston Churchill
United Kingdom Prime
Minister during WWII

After graduating from Purdue University in 1977, I hired on at General Dynamics, Electric Boat Division in Groton, CT. My assignment was to be part of a naval architecture team working on the Trident Submarine. Structural analysis fundamental principles apply to submarines as well as aircraft. Well, a contract dispute with the U.S. Government triggered massive layoffs and my career in the submarine business lasted only five months.

But I was fortunate to be offered a transfer to the Convair Division in San Diego. For the next couple of years, I worked as a Structural Engineer / Stress Analyst on the Air Launched Cruise Missile (ALCM) Program. Soon, I also was working on my Masters Degree in Aerospace Engineering at San Diego State University.

I highly recommend a Masters Degree for any engineer who wants to follow a Technical or Manage-

ment career path. It will help build your technical foundation by providing you with deeper technical knowledge and understanding associated with complex challenges and problems, and perhaps result in a higher level of performance. The field of engineering is so competitive, an advanced degree and a concurrent higher level of performance are some of the few ways to differentiate you from peers. Further, I suggest you pursue this degree while you are working to enhance the context of what you are learning.

After two-plus years on the job, I really wanted to work on military aircraft and applied for a job with McDonnell Douglas in St. Louis. My wife and I both had roots in the Midwest, and we wanted to be geographically closer to our family while planning to start one of our own. My employment began the Monday after the Thanksgiving holiday in 1979. I was able to complete the last few weeks of my masters program virtually in December, then receive my diploma the following June. As with my Bachelor's Degree, my emphasis of study was in Structures and Structural Analysis, and also included course work in Vertical Short Takeoff/Landing (V/STOL) aircraft, rotorcraft and dynamics, all which contributed to building my technical foundation.

It was a great time to join McDonnell Douglas in St. Louis. The last of more than 5,000 F-4 *Phantoms* were still being built (to join those already delivered to the U.S. Air Force, Navy, and Marine Corps, and many international customers). F-15 *Eagles* were now in production, with many already delivered to the U.S. Air Force. And the F/A-18 *Hornet* and AV-8B *Harrier II* were just then in development.

Basic Primer: How Aircraft Fly

As the examples presented throughout this book all involve aircraft programs, I'm including some definitions and a brief explanation of "how aircraft fly."

Using a fighter aircraft as an example, the basic components or major assemblies of a fixed-wing Aircraft are a fuselage, wing, and empennage (the vertical tail or stabilizer and rudder, and the horizontal tail or stabilizer). Wings include control surfaces such as flaps and ailerons. When the components or major assemblies are physically joined together, you have an "airframe." Add avionics and software, a flight control system, landing gear, engines, canopies, ejection seats, fuel and other electrical and mechanical systems, and you have an air vehicle. Add radar, defensive systems, weapons, mission avionics, and a Universal Aerial Refueling Receptacle Slipway Installation (UARRSI) or fuel probe, and you have a complex military aircraft or weapon system. See Figure 4.1.

There are four forces always acting on an aircraft in flight. An aircraft's wing camber generates LIFT as airflow passes on top of and below the wing pulling it up. The aerodynamic forces offset the WEIGHT of the aircraft and enable the aircraft to maintain altitude. DRAG or resistance is created by the aircraft as it flies through the air. The engines produce THRUST that overcome the drag and enable the aircraft to maintain constant speed, or accelerate based on pilot command. See Figure 4.2.

A "flight control system" allows the pilot to maneuver, to create roll, pitch or yaw by sliding the control stick left or right, pulling the stick back and forth, and pushing the left and right rudder pedals. Ailerons on the wing are the primary control surface

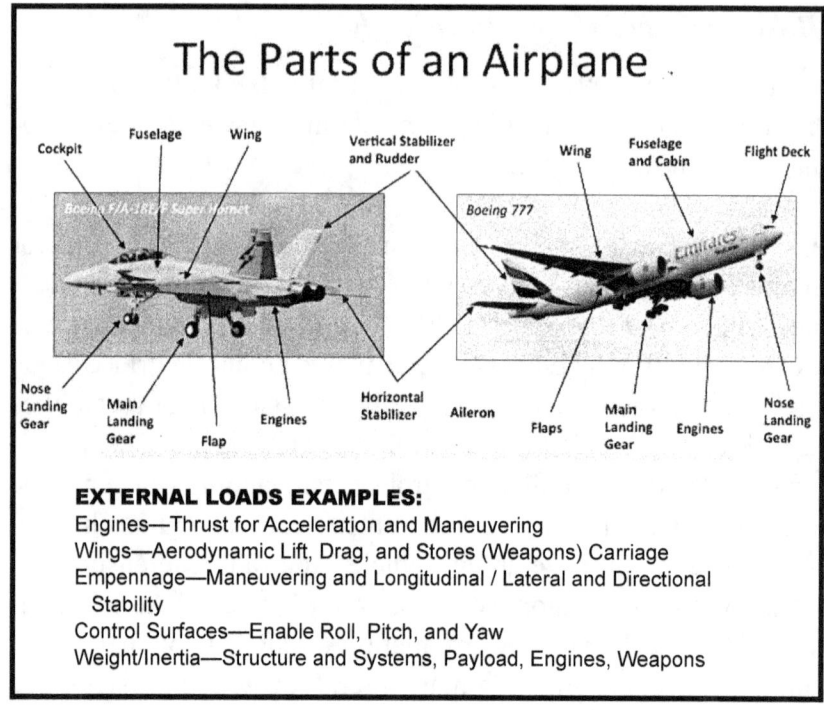

The Parts of an Airplane

Cockpit
Fuselage
Wing
Vertical Stabilizer and Rudder
Wing
Fuselage and Cabin
Flight Deck

Boeing F/A-18E/F Super Hornet
Boeing 777

Nose Landing Gear
Main Landing Gear
Engines
Flap
Horizontal Stabilizer
Aileron
Flaps
Main Landing Gear
Engines
Nose Landing Gear

EXTERNAL LOADS EXAMPLES:
Engines—Thrust for Acceleration and Maneuvering
Wings—Aerodynamic Lift, Drag, and Stores (Weapons) Carriage
Empennage—Maneuvering and Longitudinal / Lateral and Directional
   Stability
Control Surfaces—Enable Roll, Pitch, and Yaw
Weight/Inertia—Structure and Systems, Payload, Engines, Weapons

Figure 4.1—The Parts of Fighter and Commercial Airplanes (Copyright, Boeing)

that enables the aircraft to roll or bank. The horizontal stabilizer controls the aircraft to pitch nose up or down and the rudders provide the aircraft to turn nose right or nose left. See Figure 4.3.

Advanced fighter aircraft have "differential stabilizers," meaning they can deflect in opposite directions enabling the aircraft to roll or bank. Flaps, which change the camber of the wings and thus increase lift at slower speeds, are largely used during take-off and landing.

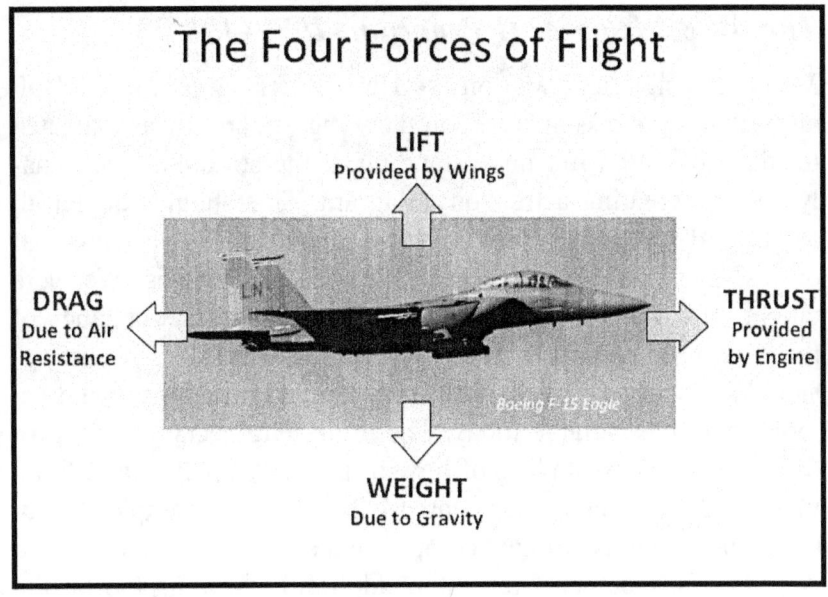

Figure 4.2—The Four Forces of Flight (Copyright, Boeing)

Figure 4.3—Front, side, and top views of an aircraft in a roll or bank, pitch, and yaw, respectively[1] (Copyright, Boeing)

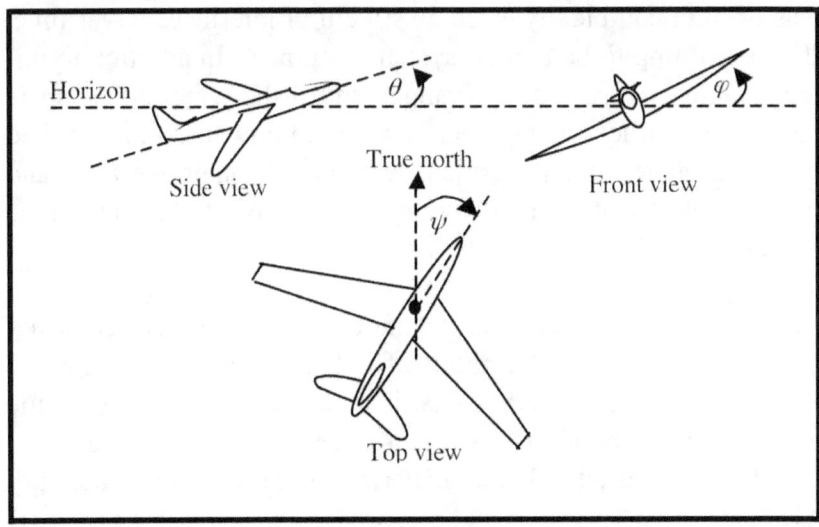

The AV-8B Harrier II Program (1979–1987)

My first assignment was on the Harrier II Program, a V/STOL aircraft, as a "stress analyst" on the wing . . . and I was expected to demonstrate a full understanding of the structure I was analyzing, by creating a freebody diagram. This should depict all forces and reactions acting on a structural part, like a wing spar or fuselage frame, prior to starting detailed analysis. We were taught how to make technical assumptions due to redundancy of the structural load paths at key interfaces to ensure a robust design. We couldn't take advantage of computers and the analytical tools that are available today. Therefore, engineers gained a lot of inherent understanding of how to properly analyze and think through the challenges associated with design of complex structures, including use of new composite materials systems. Toward the end of my career, I felt the pendulum had swung too far the other way—engineers were too reliant on computers.

Technical leads and first level supervisors advised, assisted and checked the work of the more junior engineers. Technical leads taught us systems level and critical thinking when analyzing areas of high complexity, such as structural interfaces. Over time, I was learning to be a good systems engineer. In addition to understanding the structural arrangement of the wing (skins, spars and ribs), a structural engineer understood how the wing attached to the fuselage, what the critical design load conditions were, and how the electrical, fuel and hydraulic systems interfaced with the structural members inside the wing.

Organizationally, loads and dynamics engineers provided structural engineers with definition of external loads, to be applied to a Finite Element model. That is a 3D representation of some component of the aircraft structure. Model inputs included applied loads, structural and material properties, and boundary conditions. Model outputs were structural deflections and internal loads. The latter represented the resultant bending moments and shear forces in the wing spars and ribs, and the biaxial and shear forces in the wing skins. The structural analyst would translate those internal loads into stress (and strain) analysis calculations and check them for the appropriate failure modes (e.g., tensile strength and compression or buckling).

In addition to static strength analysis, stress analysts perform checks against fatigue on metallic components, which are a result of putting many years of takeoff and landing cycles, and maneuvering load factors, also known as "g's," onto an aircraft. In the case of military fighter aircraft, these thousands of cycles of loads on wings, fuselages, empennages, landing gears, and other components will take it's toll on aircraft structure and can ultimately cause cracks. Another source of fatigue on military aircraft empennage structure are buffet loads, high cycle, vibration-like loads associated with unsteady aerodynamic flow fields from wings or aircraft with wing leading edge extensions (LEX) flying at high angles of attack. Buffet loads, if not properly accounted for in the design phase, can adversely affect fatigue life of the structure.

During the design process, customers and contractors work together to develop a "fatigue spectrum", which includes the number of occurrences that an aircraft will pull 7gs, 6gs, etc... over the course of its lifetime. The spectrum also defines aircraft configuration and considers the types of aircraft maneuvers, such as "symmetrical pull-ups" or "rolling pullouts." In addition to static testing, full-scale airframes are fatigue tested to validate structural integrity. However, once aircraft are fielded, sometimes the actual spectrum doesn't match the predicted spectrum, which can cause fatigue cracks earlier that expected in aircraft structure. The A-10 Re-Wing Case Study in Chapter Ten describes this situation in detail and how it was handled.

One of the most vulnerable areas that can be susceptible to cracks is around structural holes; especially holes in areas of high local load introduction. An example of this can be a landing gear attachment to a fuselage or a pylon interface into a wing. The interface loads at the aforementioned examples can be very high during takeoff, landing and flight maneuvering. In addition, design details such as sharp corners or tight radii in detail parts are sources of high stress concentration. On the AV-8B program, as an example, fatigue analysis was conducted to preclude crack initiation.

The primary recipients of structural analysis were the design engineers, who would draw or model the structural and systems layouts, then the detail parts. We worked closely together to ensure that what was being analyzed could work, physically and functionally, and work right the first time. Otherwise . . . expensive and program-delaying rework occurs. I emphasize, frequently and perhaps to the point of "irritation," that teamwork is vital at every stage, and not just involving the teams mentioned . . . every element brings its own collection of teams, and the work of all must fit together.

When it came time to plan for full-scale static and fatigue testing, (with test & evaluation engineers), the structural analysts knew where to place instrumentation in order to validate the Finite Element model, which was the source of internal loads used for structural analysis. The static and fatigue test articles were full-scale airframes—there were no systems or engines installed. The objective was to apply aerodynamic and inertia loads that represented the maximum load conditions the aircraft would see in-flight that put the highest stresses on the aircraft. For a structural engineer, these tests were the verification of the team's collective efforts since development began. Specifically, this meant that the "as-designed-as-built" airframe met the structural requirements of the aircraft.

Requirements for Structural Design Loads: Definitions and Criteria

● Limit Loads are loads that represent the maximum flight conditions based on aerodynamic and inertia loads.

● Ultimate Loads are Limit Loads with a Factor of Safety: Ultimate Loads = Limit Loads x 1.50.

● The 1.50 "Factor of Safety" is applied to account for differences between actual and predicted loads, structural analysis computations, variation in material properties, variation in the manufacturing process, and over-stress during aircraft operations.

● The Factor of Safety, 1.5, is an industry standard for design, analyses, and test of aircraft structure.

● Fatigue Stress Analysis is generally documented against a stress allowable that represents four (4) Lifetimes of a spectrum that represents operational use. A Full-Scale Fatigue airframe will be tested to two (2) lifetimes. In the P-8A Fatigue Test Case Study in Chapter Ten, the customer planned ahead and is conducting three (3) lifetimes of testing for purposes of life extension.

During static testing, it was exciting—call it, "nerve racking"—to watch the tests, as a failure below 150 percent of the maximum or Design Limit Load conditions would be considered as a "no-pass." That would trigger some level of redesign and be a setback for the program. A small area of the Harrier wing static test failed at 183 percent; this was well above the 150 percent limit but nonetheless was a cause of some concern until we understood the failure mode and because the test was run at room temperature. When "adjusted" for real-world operational conditions, all was well and within specification.

Once full scale ground testing was completed, Boeing and U.S. Marine Corps test pilots would test the aircraft in flight, taking it to the extremes of the operational envelope. I can still remember early in 1985 the two U.S.M.C. pilots, who had completed "OP-EVAL" or operational evaluation, at Cold Lake, Canada, China Lake, CA. and at Yuma, AZ., came for a visit. They stood on cafeteria tables and told us how well the test aircraft had performed in an operational environment. Made me proud to be part of the *Harrier* program team.

About three years into my assignment, I was asked to rotate into the Loads and Structural Dynamics team. At first, I thought I must have done something wrong . . . "job rotations" were not common. I was assured, this was a great opportunity, greatly adding to my knowledge base. Today, such rotation is common.

With looking for advancement and the opportunity to obtain more responsibility, this rotation provided some differentiation from my peers. Up to this point I was just another member of the pack, getting raises more or less on schedule just like almost everyone else. I seemed to be advancing at some sort of "programmed" pace.

I was taking advantage of a Voluntary Improvement Program. With after-work classes taught by senior level engineers who had wide-ranging expertise in multiple subject areas, this was an excellent and very convenient way to help build my technical foundation. McDonnell Aircraft and the Astronautics Companies were filled with engineering experts who offered classes in Fatigue and Fracture Mechanics, Structural Analysis Best Practices, and Flight Mechanics / Stability and Control. I was growing in knowledge . . . and confidence.

In the mid-1980s I had the opportunity to be part of a pilot program, which was defined as a "collocated, multidisciplinary team that implemented an Integrated Product Definition process (IPD)." Our task: to design and produce a unique nose cone for the United Kingdom's Royal Air Force (RAF) Harrier which could withstand a collision from a bird larger than the specifications for the then-current U.S.M.C. design.

> *I saw first hand the damage that a bird could do to an aircraft in flight. A couple of years prior to this effort, a U.S.M.C. Harrier took a bird strike on the nose, which penetrated the structure. Both the nose cone and front end of the forward fuselage required structural repair. I was asked to be part of a travel team (with three mechanics, an inspector, a shop supervisor and another engineer) to Marine Corps Air Station (MCAS) Cherry Point—the same place I first watched fighter jets as young camper.*

The revised structure was made from graphite epoxy reinforced with Kevlar fiber to enhance energy absorption. It was a real opportunity for engineers of different disciplines to work in an integrated fashion.

Designs were being reviewed and manufacturing engineers and tool design engineers provided feedback. Input from the factory floor was incorporated prior to engineering release. The people in these two disciplines were the engineers who collectively established the work sequence and tooling for the build process.

The opportunity to work together with people I wasn't familiar with was a real learning experience.

Collocation facilitated real time communication, alignment and integration of people with team goals. A key difference working in this environment vs. "business as usual" was that the cost profiles shifted more man-hours to earlier in development. These were predominately for manufacturing engineers and tool designers, who provided real time feedback to the design team. Manufacturing work instructions and tool orders would be created concurrently before release to part fabrication and to production for assembly. Traditionally, these two functions would receive the engineering drawings after they were completed and released. As a result, there often were missed opportunities to provide the production or build team with the most efficient build sequence and clarity in work instructions to assemble the aircraft. The IPD process also provided better part dimensional quality and tooling concepts that would enable parts to achieve form and fit requirements the first time without the need for design changes or rework.

The IPD process results in a lower first-unit cost, depicted as "T-1" in Figure 4.4, and a flatter learning curve for the production team.[2] The basic concept of a learning curve: the cost, of performing a task (e.g., producing a unit of output), decreases at a constant rate (expressed as the learning curve percentage, for example—85 percent) as cumulative output doubles. The savings over the production run is the area shown between the two curves.

Prototype testing was conducted, as we had modified the composite material system. The test plan called for firing representative (already dead) birds at structural test specimens to make sure the design requirement of "no bird penetration" into the structure

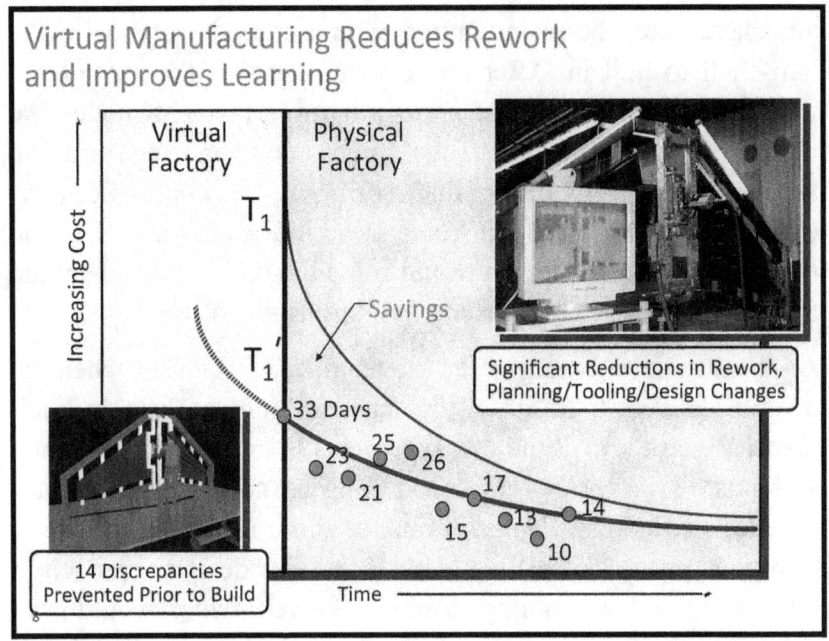

Figure 4.4: Production Learning Curve Benefits from the Integrated Product Definition Process.[3] (Copyright, Boeing)

was met. While the actual nose cone was designed as two conical shaped halves, we fabricated half cylindrical shell test specimens that represented several cross sectional cuts (skin radius, laminate thickness and frame configuration). After autoclave (an oven that provides heat and pressure to a composite laminate) cure, the parts were removed from the fabrication tool. The test specimens built as half cylinders "sprung" inward, essentially changing the radius of the parts. It was an acceptable practice to apply light hand pressure to "slightly" open the part to meet the contour, but some of the parts were too thick and stiff. We were surprised, but fortunate to find the problem when we did.

The root cause was attributed to building thick half cylindrical shells with a fairly small radius. During part cool-down after the autoclave cure, the resin contracted and caused the ends of the half-shell to pull in. After the team understood the root cause, which we validated with a structural finite element model, we reached out to some U.S. Government engineers we heard had similar challenges with "laminate spring-in." A couple of us traveled to then McClellan Air Force Base in Sacramento, CA, and through their feedback, we would solve the problem by changing the laminate design to preclude the "spring-in" of the parts.

As the structural engineer for the team, I was present when the dead birds were fired out of the "chicken gun" at the test specimens. When they hit the test specimens there was visible blood and guts. It was, however, a good thing we conducted the prototype tests as (1) it validated or anchored our analysis to preclude bird penetration and (2) we learned that our design approach to fabricating and assembling parts was flawed, which would have led to a lot of costly rework with impact to program schedule.

There are several best practices to highlight:

1. Bird strike analysis was not an easy problem to design, as there were many variables; which included;

a. The angle of impact varied with the changing contour, or angle of incidence from the nose of the airplane back to the forward fuselage interface.

b. Skin thickness and substructure or frame sizing, which provided foundational support to the skin

c. Estimating the equivalent loads of a bird impact at the design velocity.

d. Because many technical assumptions were made, and given the experience or lack thereof on bird strike analysis in the 1980s, testing was done to validate our assumptions and anchor our analysis.

2. Through structural modeling and analysis, we were able to understand the root cause of the spring-in phenomenon before taking any action.

3. We were willing to learn from others who had experienced a similar problem, and meet with them face-to-face to understand their approach to problem resolution.

4. We applied technology and an innovative approach to solving the problem. Instead of a costly redesign that would have affected the build process, we took what could have been a major setback, and instead, solved the problem quickly with a simple laminate change.

The multidisciplinary team approach worked well. Design problems were addressed real time and the necessary changes completed without adversely impacting production and delivery of parts to the production line. This was an excellent example of how teams should work together as well and taking the initiative to learn from others. Further, decision-making was pushed down to the responsible people. While the team approach enabled a more nimble response to what would be characterized as a small development project, the team exemplified "first time quality" engineering. Meaning, while we had challenges during development, and all programs had them to one degree or another, the problems were identified at the right time in the development process as to not "travel risk" to the build process.

My definition of "traveled risk" is *engineering work that is not completed or not completed right during the planned phase of development.*

In the case of the RAF nose cone, had the engineering drawings been released to the fabrication team with the original skin laminate design, the production parts would have been unusable. This would have caused major disruption to the *Harrier* production line, requiring a redesign and manufacture of new replacement parts in the middle of the production phase of the program. This would have been costly and impacted the program schedule. In this case, through our prototype test specimens built for bird penetration testing, we found and corrected the problem before engineering was released, thereby avoiding what would have been cost overruns and schedule delays. The collocated, multidisciplinary team structure was a very effective team structure and has since evolved into Integrated Product Teams (Chapter Seven).

Business Acumen

I was selected to attend an Executive-Master of Business Administration (MBA) Program at Washington University, which I started in 1992. This two-year program provided me with much needed business skills, especially in analyzing data and turning data into information for making decisions. This gave me the foundation and tools to become an executive leader.

If you are in corporate America, you work for a company that is in business to provide products and services to customers and make money in the process. Orders, revenue, earnings, cash and stock price are all financial measures that indicate the health of a publicly traded company. It's important for engineers, and just not those in leadership positions, to have business acumen skills and understand the relationship of the cost of doing business and how it relates to the top and bottom line of the company. As engineers, our role in the company is to design, build, test and support products; we are expected to complete our tasks, on time and within the budget. There are many more dimensions to business, which can be learned, both on the job and in school. One asset: Boeing's Leadership Center provides outstanding leadership training and business acumen, to people throughout all levels of management.

Through my MBA Program, I learned how to think like a senior executive: how to read an annual report, develop and execute strategies that beat competitors, and adapt to changes in corporate direction associated with the global environment. I learned "best practices" in human resources and the importance of "culture" when conducting international business. You might call these topics and concepts "the tip of the iceberg" in a crowded program. Going back to school while holding down a full time, responsible and demanding job was truly a challenge. With the support of my family, I was able to successfully complete the program with Honors in May 1994.

Frequently, early and mid-career engineers would ask me about the importance of an MBA and wonder, when was the best time to pursue? My advice: get an advanced technical degree first, perhaps in early-career. If they decide to take a career path into management, perhaps somewhere between eight and fifteen years as a working engineer, then was the time to start an MBA program. Based on my personal experience, I felt the engineers would take away much more from an MBA after they had worked in industry for several years. By having a broader range of work experiences at this point in their career, people would be able to (1) better understand what they were learning and (2) more effectively apply it to their current and future job assignments. ■

Chapter Four Technical Takeaways

● Computers are a useful part of the design process but not a substitute for gaining a basic understanding of the design and critical thinking.

● Building a strong technical foundation, specifically breadth and depth of knowledge, is important for you to be effective as an engineer and prepare you for future growth opportunities.

● Learn early in your career how to be a good systems engineer in your area of expertise. Strong systems engineers have a comprehensive understanding of the product being developed, built, tested and supported. Having this knowledge is key to achieving a first time quality design.

● Guard against too much reliance on computers and computational power to drive analytical conclusions and technical decision-making.

● Anchor technical assumptions with analysis and test data. Consider rapid prototyping to prove out or validate key build processes if methods are not well documented or understood.

Chapter Four General Takeaways

● Job rotations are an excellent opportunity to broaden

your knowledge base. They can also separate you from peers, both at salary/performance review and for future job opportunities.

● Keep learning. Take advantage of continuing education opportunities within and outside your company.

● Take charge of your own career. Seek out advice from your leadership and mentors who can provide you with career guidance based on your aspirations.

● Business acumen is an important skillset for engineers, especially those with executive leadership ambition. But . . . spend time building your technical foundation before pursuing an MBA degree.■

---

1. Lavretsky, Eugene and Wise, Kevin, Robust and Adaptive Control With Aerospace Applications, Chapter 1, 2013

2. Martin, James R. Ph.D., CMA Learning and Experience Curves: http://maaw.info/LearningCurvesMain.htm Retrieved on 12 NOV 2015

3. The curves plotted on Figure 4.4 are normalized and based on a combination of (out of) production program data and use of assembly simulation tools.

CHAPTER FIVE
EARLY TO MID-CAREER TRANSITION

If you have a positive attitude and constantly strive to give your best effort, eventually you will overcome your immediate problems and find you are ready for greater challenges.

Pat Riley
NBA Basketball Coach

In January, 1989, I was home on a Friday night having dinner with my family. I had been working on another development program for about 18 months. The phone rang and my boss called; this was a very rare occurrence back then, at least for people at my level. The message: his boss directed that I report back to the *Harrier* Program. There was a meeting the next day, a Saturday morning, when I would learn about my new assignment. That phone call would change my career path in a way I would have never imagined.

AV-8B Harrier Wing Team Leader—1989

I arrived early the next morning, maybe even the first one to arrive, as I was anxious to see what was going on. I had been promoted about six months earlier to a first level leader on the YF-23 Wing Structural Analysis team. So . . . what was next?

Soon, the room was filled with executives within and outside of the program, most of whom I had very little contact with if I knew them at all. They were talking about a significant issue on the Harrier wing; many wings in the production flow were found to have significant "delaminations" in the composite wing skins. Delaminations are cracks between plies of the skin and were found and characterized (location and size) with use of "non-destructive" or ultrasonic inspection techniques. We met until late in the day and then again on Sunday. I recall the weekend meetings as being very intense. Program managers and other executives wanted to know what went wrong and how was it that so many wings were built before the problem was identified? There were hard questions and no one had good answers.

Basic Primer on Composite Materials

Composite material skins were made up of individual plies, combining unidirectional carbon fibers with an epoxy resin. The strength and stiffness of each of these two constituent materials by themselves is negligible. However, when you combine the carbon fibers and epoxy resin together, then apply pressure and temperature in an autoclave, the strength and stiffness are almost comparable with aluminum—but with a density about half that of aluminum—which meant lighter weight. Lighter weight improved vertical takeoff and landing performance. Also, a composite skin laminate made up of plies with fibers at different orientations provides consistent load-carrying capability under varying aerodynamic and inertial forces. Think of a truss—as used in construction of bridges and buildings, having horizontal, vertical and diagonal structural components. However . . . composite materials must be designed and constructed in a way that doesn't put high levels of transverse-shear stress between the plies of the laminate.

The *Harrier* wing delaminations affected structural integrity. Now, since each *Harrier* delivered had flown a maximum load factor condition during acceptance flights, the possibility of near-term failure seemed remote enough that the fleet did not need to be grounded. The Naval Air Systems Command (NAVAIR) Structures Division set flight restrictions for operational aircraft to about two-thirds of its design envelope speed; no new aircraft would be delivered until the problem was resolved. Every aircraft in the fleet was to be inspected for delaminations, As you will appreciate, halting deliveries had a huge impact on the program—emotional, and financial. My new job: to lead a team to (1) fix the build process, which meant providing objective evidence that the manufacturing process was in control, (2) to inspect, discover and repair the airplanes in production that had wing delaminations, and (3) to coordinate inspection of fleet aircraft to ensure there were no wings with delaminations. We had to meet all three objectives to convince our customers—and ourselves—it was safe to operate the aircraft across the full flight envelope.

With help from senior management, we staffed a multidisciplinary team within days with the right skilled engineers.

This would be the biggest-yet test of my leadership, as I would be leading a multidisciplinary team which included engineering, manufacturing and quality skills; people that normally didn't work together on a daily basis. I felt technically prepared: I knew the guts of the wing design, the implication of the delaminations, and the process of dispositioning the non-conforming conditions.

What I didn't fully appreciate at the time was how a production system could get out of whack so quickly, and the impact on our customers. To say the least, the customers were very unhappy at every level.

I also learned very quickly that the team didn't have a comprehensive understanding of the wing build process, including the
sequence of assembly and the tooling. Clearly we would have to
learn, and soon, if we were going to define and implement effective corrective actions. By being collocated on the factory floor
side-by-side with the mechanics, my team learned how the build
team assembled the wing, and discovered what they believed to
be some of the driving root causes. It was not, however, easy.

The days—12-hour days, six, sometimes seven a week—started at 6:00 A.M. with a review of the day before, the latest reports of customer feedback, and a plan for the new day. Initially,
there was a lot of finger pointing, people trying to blame each
other, and things would get worse before getting better. Emo-
tions were running high throughout the teams and at all levels of
management.

The pressure was intense; every day without a solution would
impact both production and the warfighters. It didn't matter how
many hours and days the team worked, our leaders wanted results. It was difficult at the outset hearing it all, but I tried not to
take it personally and to stay focused.

It was clear we needed better teamwork. We were so busy working to correct the problems, there would not be any formal team-
building events; at least for a while. My approach was to get the
whole team aligned and focused around the task at hand.

We instituted a Root Cause and Corrective Action process
(RCCA). Understanding root cause was imperative—we couldn't
know what to do to fix the problems until we understood how
they happened. When did the fabrication and or assembly

processes change? The team spent a lot of time on the factory floor with the people who built and performed in-process inspections to get their inputs, which were invaluable. The mechanics and technicians helped the engineers also understand the build process, and the tools and tooling used to assemble the wing. Potential root causes were documented, action items assigned and followed up daily.

The Manufacturing and Quality teams are an excellent source of information when solving continuous improvement opportunities that impact productivity. Seek out their advice and make them part of the solution.

We uncovered key "root causes:" Substructure parts (that is, parts that are inside the wing) were not always installed flush with the wing skins. Whether the parts were out of dimensional specification or not properly located in the wing, the result would be gaps between the substructure and skins.

1. Tools used to position these parts lacked features that produced a repeatable process given the design tolerances.

2. If these gaps were over a certain limit and not properly filled with spacers or shims, the torque associated with fastener installation would squeeze the composite wing skin to a point that it caused the delaminations.

3. Most significant root causes: the manner in which "gap checks" were performed, and the installation (or lack thereof) of localized shims. We discovered that the process of gaps checks had been changed at some point before the problems began.

4. There were other contributors such as: fastener

torque guns were not properly calibrated and sub-assembly tools had to be changed, to improve substructure-to-skin alignment.

Concurrent with the root cause analysis, we had to "contain" the problem. We had 26 wings in-flow that had to be inspected for delaminations and fixed, as appropriate.

1. A big challenge was to determine if and when build process changes occurred that could have caused these delaminations. We searched the data and saw that there was a point in time where there was a step function change in quantity of shims in each wing installation.

2. As a result, wings before and after the time of the process change were inspected. Through inspection of a statistically valid sample of wings with non-destructive inspection methods, the team validated the hypothesis.

3. Once we bounded or contained the problem, the team processed every repair on every wing that had delaminations through a disciplined Material Review Board process. It was painful and time consuming. The engineers on the team were outstanding, working to determine the most effective and efficient way to disposition every wing. They documented their analysis and presented every page of analysis to our customer for technical concurrence.

The next step: to implement "Corrective Action."

1. Our NAVAIR Structures customer alerted us that the

A-6 Re-wing Program at Boeing-Wichita was having similar delamination problems. The A-6 Re-wing Program had wing skins also made of composite materials. We arranged a benchmarking visit to specifically understand their problems and how they were planning to correct them. Their problems were similar in that they were also experiencing abrupt, unshimmed gaps in the substructure, or parts beneath and between the wing skins. These wing parts are also called spars and ribs.

2. Their approach was to apply a plastic "liquid-shim" material across the entire wing skin-substructure interface. This shim material was a liquid plastic, which hardened and filled the gaps. We discussed their experience and process on how to successfully apply the right amount of shim and structural limitations on thickness of shim material.

3. The team built test specimens and applied loads in a manner that represented the wing skin-to-substructure part interfaces. This enabled the team to quantify the reduction in strength for parts that were mechanically fastened together with liquid shim material. We did this because we didn't know to what degree this new liquid shim material could or would affect our structural allowables of mechanically fastened wing skin-to-substructure joints.

4. Once we tested an agreed upon number of specimens, we were able to assure ourselves that the wing would remain structurally sound with the liquid shim material.

5. With input from the team, I would make the biggest decision of my career, up to that point in time. I told leaders in Program Management, Engineering,

Operations and Quality, that this was the right way going forward to ensure we would have no more delaminations, and, that once we perfected the process we would be able to restart and maintain continuous production flow. They were "all-in."

This is a prime example of the job of a chief engineer and the kinds of decisions we make.

6. Our trial run with the first wing went better than expected but the process was slow and labor intensive. If shim thicknesses were too thick (because of inadequate or non-uniform pressure applied across the skin-substructure interface), we had to rework it.

7. Communication with engineers and the build team were constantly on-going, and important to perfect the process and make it production ready. We would pull in our manufacturing technology team to help with better tools and equipment.

8. As we worked to develop a repeatable liquid shimming process, we attacked the dimensional quality of the parts within the wing with our fabrication centers and our suppliers. Our focus here was to "minimize detail part variation," which would reduce or eliminate gaps altogether.

9. Further, actions were taken to put in place a more robust calibration process for our fastener installation guns, and sharpen and replace worn drill bits more frequently.

It took months—including building ten wings free of delaminations—to prove that the RCCA process was production ready. Our monthly visits to the U.S. Government technical community were, well, difficult, especially early on, while we struggled to get our arms around the root of the problem and develop a build process fix. As we made progress, the meetings got a bit easier. I should note, we were briefing not only our own leadership and the U.S.M.C. customer, but our three international customers—U.K., Spain, and Italy—also. This was the first assignment in which I had to stand up in front of management and customers, to brief on the technical issues and the steps we were taking to fix them. I must admit, at first, this was a personal challenge but things got easier as I gained confidence. Confidence, that is, not only in my own "presentation" skills, but also in the quality and accuracy of information I was able to deliver.

We were under constant pressure—move faster, solve the problems NOW—but I'm proud to say my team stayed laser-focused until we got the job done. As unit costs and productivity improved and the process stabilized, the work environment got better. The stakeholders would all see a collective path forward out of this difficult situation. About nine months (and $7 million) later, we were building defect-free wings. The customer began accepting deliveries, removed the flight restrictions, and the wings were back in continuous production. Best of all, the team restored customer confidence.

I was really proud of the team. We had grown together having been through a really tough year of hard work. Team members learned to appreciate what skills and capabilities each other had to bring to the team. It was educational for the engineering, quality and manufacturing teams to learn to better work together. This was one of the highest performing teams I would ever be associated with. As for teambuilding, once we were a few months into problem resolution, we did have some fun bowling nights;

"Shop" vs. "Engineering." If I recall, the manufacturing and in-spection team had the best scores and had bragging rights.

For me personally, the fact that the *Harrier* continues to fly il-lustrates how my decisions as the leader of this team during this crisis situation had a positive impact on this awesome weapon system, shown in Figure 5.1, which plans to fly until . . . 2025.

I grew as a leader and obtained an incredible amount of knowl-edge and insight on the manufacturing and quality side of the business. This experience had boosted my mental toughness and confidence to lead people, drive daily execution, lead change, make hard decisions, and effectively communicate and demon-strate results. Further, I began to build relationships with custom-ers, to understand customer expectations and how best to work with customers in difficult situations. In this case, the customers included Naval Air Systems Command's Structures Branch, the local Defense Contract Management Agency (DCMA), and the warfighters—U.S.M.C. and international customers.

I appreciated the assistance from senior executive leadership; an immense help along the way to our entire team. Success would not have been possible without their support. They became role models for how I led and engaged with people and teams in the future.

As I look back, engineering, as part of the Product Definition teams, could have done more to prevent this problem, such as by better managing supply chain dimensional variation and by putting in place more robust processes for build of the wing. For example: how to perform wing skin-substructure gap checks; how to shim gaps; better training in hole-drilling and fastener installation.

Figure 5.1—AV-8B Harrier II Aircraft Hovering close to ground (Copyright, Boeing)

This experience really brought home the importance of quality Engineering, including drawings, work instructions, tools and tooling, and process specifications, how all of these drive factory floor performance and enable continuous production flow. It also highlighted the need for quality control practices to minimize product and process variation. I shared this experience with my engineers, when appropriate, and set the expectation to apply lessons learned.

Best Practices and Lessons Learned . . . for Engineers transitioning into a management or first level leader assignment (As a result of my *Harrier* wing experience):

1. Set expectations and drive daily execution for your team. Everyone needs to know what must be accomplished every day. You must also ensure that technical lead engineers are in place, that all work is performed

with proven and mature processes and tools, and Lean+ engineering principles are being used. (These include: (1) prioritize work, (2) focus and finish tasks, and (3) eliminate multitasking.)

2. Learn what's important to the people on your team and what motivates them. Leverage each team members' strengths to maximize team productivity.

3. Understand the guts of your business, which means your product, where your product fits in to the bigger picture, and what are the processes and tools used to accomplish tasks. Make sure team members have all the information or data they need to start and finish assignments.

4. Customers want what they have contracted for and they want it when promised. Communicate often and frequently with them—when things are going well and not so well. In other words, build relationships when times are good so you can leverage when times are tough.

5. When faced with a design or production problem, use a disciplined root cause corrective action process to understand and correct the problem.

6. Have an inclusive work environment, where people feel that their voices, ideas, suggestions and concerns are being heard and considered. Where people are encouraged to ask for help and know it is okay to bring bad news forward.

7. Use a balance of leading and lagging indicators or metrics to measure performance to plan. These indicators can provide you with situational

awareness of the health of the program and enable the team to make adjustments if off plan. The case studies in Chapter Ten and the Glossary provide examples of leading indicators of performance.

8. First level leaders have a tough job, maybe the toughest being, having to lead your team and answer to senior management. They should not be afraid to ask for help. Teams that ask for help are the most successful.

I consider the *Harrier* wing assignment, "The job that changed my career." Why?

Because of our success, I would move into diverse assignments with increasing responsibility. In 1996—six years and several jobs later—I would be promoted into the executive ranks of the company, leading Information Systems for all of McDonnell Douglas Aerospace. Those six years (1990-1996) would be a very tumultuous time with successes and setbacks, opportunities and challenges, and organizational and leadership changes for the McDonnell Douglas Aerospace team.

The A-12 program was cancelled in 1991, triggering a significant number of layoffs. Teamed with Northrop, we would lose the YF-23 Program to the Lockheed-Boeing team. The McDonnell Douglas Automation Company would no longer exist and Network management, mainframe and desktop computing would be outsourced to IBM. There would be technical challenges in Long Beach, CA, on the C-17 and MD-11 aircraft programs. And, oh by the way . . . there was a 100-day labor union strike in St. Louis through the summer of 1996

On the flipside, the T-45 (U.S. Navy training jet program) would move from Long Beach, CA, to St. Louis. We won orders to build 72 F-15E type aircraft to Saudi Arabia (F-15S) and 25 aircraft to Israel (F-15I). Both variants would include some development, per direction of the U.S. Government. The U.S. Air Force ordered 17 additional F-15Es, which would be their last purchase of this venerable fighter aircraft. The F/A-18 *Super Hornet* (F-18 E/F) would begin development..

As legendary Green Bay Packers Head Football Coach Vince Lombardi said, "Luck is what happens when preparation meets opportunity."[1] Indeed; two key attributes to successful job performance are preparation, and the ability to anticipate.

Let me highlight some of my mid-career assignments in that timeframe and key lessons learned from each.

Quality Processes Division

After my experience on the Harrier wing, I was selected for a senior management position in the quality organization—an element of the newly formed Quality Processes Division (QPD). The primary focus: to develop and improve business and technical processes critically important to McDonnell Douglas Aerospace.

As I have noted, back in the day, it was a rare but big deal to move from one department of engineering to another. However, moving outside of engineering to another function, like Quality, was looked on as risky, but I was willing to take that chance. To be clear, I wasn't sure there would be a good place for me if I wanted to come back to engineering. I felt confident in my ability to perform so I took the chance. It was yet another opportunity to grow my technical foundation, strengthen my leadership skills, and build relationships with leaders across the company. As it turned out, moving back into engineering would not be a problem.

The Quality organization had been realigned to add more focus on preventing quality escapes, rather than inspecting during the build process. This was really a great opportunity to apply what I had learned in my last assignment and apply across the enterprise.

I learned there are three components of what is known as the "Cost of Quality": (1) prevention, (2) appraisal, and (3) failure. The underlying causes were across the spectrum from engineering design to workmanship.

Prevention cost has many dimensions. One of the larger comes from poor engineering quality. We formed multidisciplinary and integrated product teams, which used three dimensional drawings (3D) engineering and virtual reality design tools. They eliminated many errors associated with parts that didn't fit right or interferences between structure and systems. Manufacturing was also taking steps to prevent nonconformances by bringing automation to key process areas such as drilling holes.

Appraisal costs—auditing of factory floor build and inspection processes—were another proactive approach, to identify continuous improvement opportunities and correct issues before they became a problem; that is, before they were to be discovered during end item inspections. The audit team within QPD took a data-driven approach to target focus areas, whether driven by engineering, supplier management and procurement, or manufacturing. The team prioritized processes that manifested themselves in Failure Costs, such as rework, repair and scrap. Findings and recommendations were documented for systematic root cause evaluation and fix. Audit closure took place after verification of corrective actions.

Appraisal costs also included in-process or end-item inspections. Factory floor work instructions included critical points in the build process that required an inspector to check and put their stamp of personal accountability affirming that the product was built per drawing requirements. As the integrated product definition processes evolved, engineers would include key characteristics, or critical dimensions, to be measured and recorded. Through the years, the Manufacturing Team had evolved to "operator verification," where mechanics and electricians were self-inspecting their own and each other's work. The Quality organization continued to perform inspections at critical points in the process.

I spent two years in this assignment, and the key lesson was the vital importance of first-time quality. You could be an engineer, a supplier manager, or a mechanic on the factory floor—it didn't matter. You had to do the job right. In our highly-competitive business, the cost of rework, repair, or scrap was too high. We had to drive waste out of the system from the beginning, not wait until some quality defect was discovered somewhere down the line.

It was during this assignment that the Department of Defense cancelled the A-12 Program and I had to make the difficult layoff decisions within the Quality organization. This was one of the hardest things I ever had to do. This brought me back to the very beginning of my career. It reaffirmed for me just how important executing current contracts were to the health of a business.

Douglas Aircraft Company began development of the MD-12, a new four engine aircraft in 1991.

McDonnell Aircraft Company would have the responsibility for esign of the wing. I was selected to lead the wing torque box IPT. Note that this was the third wing assignment in my career. This program would only last about 15 months or so, before the company 'pulled the plug' on development. Key lessons learned for me included:

- Paraphrasing what GE Chairman and CEO Jack Welch said, if you were not #1 or #2 in the market, it would be very difficult to stay in the business to develop and sell product, in this case commercial aircraft.[2] With Boeing and Airbus dominating the commercial aircraft market, there was no room for us.

- During this 15-month assignment, I learned a lot about the early phases of a development program. Such as, the need for robust configuration trade studies (which considered design for manufacturing and assembly), and defining and managing major interfaces across the aircraft. I would also learn the differences between commercial and military aircraft during development, such as external and internal loads development processes for both design and airworthiness certification. Perhaps the biggest lesson learned was the difference in the cultures between geographical sites; something I would learn and experience much more of in the future.

Two factory support assignments would further broaden my skills in manufacturing and quality; (1) on the T-45 aircraft, which had been transitioned from Long Beach to St. Louis and (2) for all production programs in all St. Louis assembly areas and hangar (or flight ramp) operations.

1. **The T-45 Goshawk** was a derivative of the British Aerospace (BAE) *Hawk*. I reported directly to the Director of Manufacturing and my assignment was to lead the engineers on the factory floor. It was not in the norm for engineering managers to report directly to a manufacturing director. It was the right thing to do for the program and frankly I learned a lot from being part of this team.

Similar to *Harrier* production, BAE provided the center / aft fuselage structure to the program for final assembly. The structural assembly came with a crate of subsystem parts (e.g., fuel and hydraulic tubes) for installation in St. Louis. This aircraft was difficult to build and thus very expensive, especially installation of subsystem parts. I distinctly recall the aircraft had "hard" system and structural interfaces, meaning the design did not incorporate any flexibility for attaching fuel and hydraulic systems to structure. It reaffirmed the importance of designing the aircraft with consideration for manufacturing. We would take some significant steps with the program to improve how we built the aircraft, to include changing BAEs statement of work to install their subsystems into the aircraft prior to delivery to St. Louis for final assembly. Building relationships with the program teams was also a key responsibility, as we would have multiple technical challenges in final assembly and hangar operations requiring technical support.

2. *Engineering Support for all Manufacturing (St. Louis).* This assignment included engineering support for the production programs F-15, F/A-18, AV-8B, and T-45, and I became part of both the Engineering and Manufacturing Leadership Teams. With the mantra of "support the mechanics on the shop floor," engineers who supported production were moved from their offices to shipside; the goal, to improve productivity by solving continuous improvement opportunities real-time and with timelier disposition of non-conforming material. At this time, the company was making changes, to realign program teams into IPTs, holding them accountable for factory floor performance. My job was to align my Factory Support team with the IPTs, and make sure they were working together. Like the *Harrier* wing, this assignment reaffirmed that the quality of engineering directly affected the efficiency of the build process. Factory floor assignments were a critical step in building not just my technical foundation, but to strengthen my knowledge of the aerospace business.

Information Systems. In January 1996. I was selected to lead Information Systems for all of McDonnell Douglas Aerospace (MDA) as a division director, which was my first executive leadership assignment. MDA included all military aircraft, missile systems and helicopters.

This responsibility included working with, and managing a contract with, IBM for computing services. Our goal: to bring strategic cost advantages, by leveraging computing technology across the programs and the functions.

While I had long been a "user" of information systems, this was a new dimension and I was excited to take it on. This would be an excellent opportunity to apply my leadership skills from my previous work experiences and my recent education.

I started talking to lots of people . . . and without really thinking about it, I was building relationships. With my team, my boss, the person that preceded me in the job, our supplier partner IBM, and our customers, who were both the functional organizations within the company, and the programs. The functional organizations were responsible for processes, which defined the requirements for new information systems. Further, we deployed information professionals to programs to leverage technology, improve business processes, and to improve productivity.

The Information Systems organization at the time was part of Finance, where I had the opportunity to work for the McDonnell Douglas Aerospace CFO and with members of the Finance Leadership Team. This included the controller, vice president of contracts and the program CFOs. It was another benefit of this assignment, a great opportunity to build more relationships, to understand what their jobs were all about and how my team could best support them. Working together with the controller every week, we would evaluate for approval / disapproval every capital and associated expense request associated with computing infrastructure across the entire MDA organization.

In the early-mid 1990s McDonnell Douglas outsourced the computing infrastructure network to IBM's Integrated Systems Solutions Company (ISSC). IBM also installed and maintained mainframe computers and provided desktop services. This was during a time when desktop computing was emerging and really

changing the way people were doing business across the corporation, the country, and globally. It was a period of growth, not just
in capability and knowledge, but in dealing with new and emerging issues . . . such as, "network stability." When the network was
down, productivity was adversely affected, and most, if not every
senior executive in the company was not happy. Not to mention,
anyone in the company who couldn't get a connection. Everyone
had my phone number, and everyone was ready to use it. What
was I going to do about this?

One key, of course, was to have a good relationship with ISSC
and in fact, their client executive's office was set right next to
mine. Another key was to ensure that we could measure performance and create corrective action plans for any metric—up-
time, down-time, service call center response-time—not meeting targets. There were challenges, and frustration, and steady
improvement. (I might mention, however, that some years after
Boeing and McDonnell Douglas became one company, Boeing
would take back control of the network, it was such a strategic
component of the business.)

To improve the efficiency of production across the company, we
created the "Integrated Manufacturing and Control System." The
primary objective: to schedule delivery of parts so that they would
arrive at the assembly point when required—not before, and not
after. The benefits were obvious; reduce production delays and
also reduce "in waiting" inventory. Or, in technical terms, "reduce the critical path cycle time." This project was managed by
a team from manufacturing and information technology—folks
who knew how the product was made, and folks who knew how
to track processes, supported by software provided by a commercial supplier. To ensure that production would not be negatively

impacted when the system was first switched on, development included some real-world real time exercises, which helped weed out potential glitches.

Key learnings from leading Information Systems:

1. Systems follow processes; hence the effectiveness of our information systems was only as good as the processes of the functional organizations. One learns a lot about the business when they manage and integrate the systems based on and between the functional organization's processes.

2. Functional and IT teams have to work together to define the "As-Is" and "To-Be" processes; then ensure the systems follow the processes prior to deployment through conference room pilot programs. This was an excellent opportunity to streamline processes and remove non-value-added tasks.

3. Managing the performance of a commercial supplier with contractual metrics was critically important. Desktop computing was emerging; employee productivity was highly dependent on our supplier's ability to manage computing network infrastructure, and to respond to and resolve questions and problems through their Help Desk. We worked together closely with our supplier to ensure contractual commitments were met and took appropriate action when they were not.

4. Other than as a user of information systems, I lacked knowledge and experience in developing new applications, which were underway in HR, finance and operations. I had to delegate and rely on my team for

technical decision-making—more than I was used to, or at first, comfortable with. "Learning to delegate" would be instrumental to success in my future jobs as a Business Unit Chief Engineer.

5. Obtaining working knowledge of information systems was vital. This would become a key area of focus during the transition period when McDonnell Douglas and Boeing would become one company.

Boeing and McDonnell Douglas Becoming One Company

On August 1st, 1997, McDonnell Douglas and Boeing became one company. The primary driving factors:

1. Boeing had a large commercial business and relatively small defense business. As a result, Boeing was impacted by the business cycles of airlines, freight companies and leasing companies who bought airplanes. The defense business base was not large enough to offset the cyclical nature of the commercial business.

2. McDonnell Douglas had a relatively large defense business base and small commercial base. By this time, EADS-Airbus was just behind Boeing in the commercial aircraft business. With the cancellation of the A-12 program and loss of the YF-23, the defense business would largely be restricted to extending existing product lines as a means of business growth. Not a happy prospect, at least, initially. Further, Boeing would benefit from the addition of McDonnell Douglas strong engineering team.

3. Through Boeing's acquisitions of Rockwell's North

American Aviation company and McDonnell Douglas, the commercial and defense businesses would be similar in size (measured by revenue) and the cyclical nature of commercial airplane sales could be mitigated through the defense and space business, and vice versa.

4. Post merger, the Defense business would realize product, process and resource synergies from the three heritage companies, which enabled a positive trajectory of top line and bottom line growth. These synergies included the ability to offer domestic and international customers a variety of weapon systems to meet their requirements. A key competitive differentiator Boeing had was it's ability to provide an extensive breadth of capabilities to customers through people deployed at sites throughout the world...exemplifying "One Boeing."

F-15 Program Engineering (1997–1999)

My first assignment as a Program Chief Engineer was on the F-15 from June 1997 till September 1999. I was now leading engineering on what I felt was the best fighter aircraft in the world. The program was developing capability upgrades in 72 F-15S (Saudi) models, 25 F-15I (Israel) models and what would be the last 17 F-15E (U.S. Air Force) models, shown as Figure 5.2. In addition, I was responsible for the shared resource pool of engineers on the AV-8B/T-45/F-15 Integrated Product Team. This IPT had been put in place a couple of years before to provide sustaining engineering (factory and fleet support) at a reduced cost to the production programs.

After several years leading organization and teams on different programs and in different functions, I returned to my wheel-house of expertise—program engineering. I had obtained a wide breadth of diverse work and leadership experiences, and relation-ships with executives throughout the company. Having the op-portunity to lead engineering for the world's best air superiority fighter was an honor and privilege. The aircraft's performance in the first Gulf War was exemplary. The F-15 was credited with 32 kills against enemy aircraft during operation Desert Storm.[3] I recall quotes on coffee mugs, "Every single enemy aircraft shot down was shot down by a fighter built by McDonnell Douglas"; credit to the F/A-18 *Hornet* as well.

> *The Gulf War example above exemplifies that the work we do matters. For decades, even generations, our company's people have worked tirelessly to provide the United States Military and Allied Forces with highly capable weapon systems that when operated by the dedicated men and women defending our country and our freedom, they will always have the edge in combat when in the fight against the enemy.*

I had joined the program about 9 months or so after the summer long labor strike in 1996. The company made the decision to con-tinue work during the strike, so the factory floor was anything but operating with a continuous production flow—even for months after the strike had ended. The F-15 was originally designed in the late 60s, so it didn't have near the "design for manufacturing and assembly" considerations as if it were being designed in the 21st century. It took too many man-hours, and under the strike and im-mediate post-strike conditions, "man hours" were at a premium. We had to find ways to build the aircraft more efficiently.

The company's vice president of manufacturing brought in an

Figure 5.2 Topside View of an F-15E *Strike Eagle* (Copyright, Boeing)

operations leader from a different site to introduce a best practice, called a "quality walk." Every day, the heads of all cognizant
functional engineering groups—structural design and analysis,
manufacturing engineering, tool design, liaison engineering, quality—together with the program manufacturing leader and myself
as the chief engineer, would visit a couple of areas in the factory. Manufacturing superintendents would identify specific areas
that were not performing well. Collectively, the team would visit
those areas and talk directly to the mechanics or electricians and
ask him or her about the challenges they faced, and what could
be improved. It might be incorrect engineering, or lack of clarity
in work instruction, even a tool that needed some level of maintenance rework or repair. With all the leaders of the teams that
owned these resources, we prioritized and addressed these continuous improvement opportunities. Over a period of 9-12 months,

we realized a 38 percent reduction in man-hours per aircraft. This was a significant improvement in unit cost in a relatively short time. Yet another example of how engineering can help drive factory floor performance.

During my tenure on the program, we completed the build of 48 of the 72 F-15 Saudi aircraft, all 25 F-15 for the Israeli Air Force and the first of the last 17 E models for the United States Air Force. One day, the F-15 Israel Program Manager reported that the customer was experiencing problems with the Programmable Armament Control System (PACS)—which was designed, tested and built by one of our suppliers, and integrated by the program engineering team. Given the mission of the F-15 aircraft as an air superiority fighter, a problem with the system that launched weapons was unacceptable.

Quickly I got up to speed on what we knew and didn't know and my team and I flew off to Israel. We met with a room full of Israeli Air Force officers. I will never forget what the ranking officer said to us that day, "The PACS is not the weak link [of the airplane], it was the missing link." That was certainly hard to hear, but, given the environment in that part of the world, I promised immediate corrective action. Before we left the country, we had tapped an avionics engineer to lead the root cause analysis and corrective action process. Further, I had placed a call into my supplier counterpart, as we had developed a good relationship during the development process. Boeing and our supplier worked together to fix the problem, quickly. While I would travel many more times to Israel in my career, I never again had to get in front of the Israeli Air Force because of a technical problem. ∎

Chapter Five Technical takeaways

• A key to your professional advancement: establish a track record of delivering results in a variety of assignments.

• Communicate frequently with your customers, both when things are going well and when they are not. When problems arise, work with a sense of urgency to correct them.

• Develop and continuously hone your presentation skills. The ability to stand up in front of leadership, customers and your team, and communicate with clarity is essential.

• Ensure the quality of engineering, which directly affects the efficiency and effectiveness of the manufacturing build processes.

• Know the guts of your business. Understand the product and critical processes required to execute the job assignment you are responsible for.

• If design or process changes are going to be made, make sure that second and third order consequences are understood before you move to implementation.

Chapter Five Leadership takeaways

• Work customer problems with a sense of urgency. Make sure that root causes are known before taking action to implement corrective actions. Further, understand the processes that broke down.

• Before you actually start a new assignment, ask, what does management want you to accomplish in the first 3 months? You might want to develop a 30-60-90

Day Plan to guide your transition.

● Don't take on too many tasks too quickly. Prioritize and focus on what's most important.

● Your performance on the job will not be assessed on the number of work-hours you log. You will be scored on "results."

● You can be a highly effective leader—even if you aren't a domain technical expert—if you surround yourself with the right people and build strong and effective teams.

● Welcome "change" and the opportunity to be in charge of the attack. Senior managers look for people with the fortitude and courage to step into tough assignments. And who deliver results.

● Your job is what you make it. In every one of my assignments, while there were clear expectations from senior leadership, I took charge, shaped the job, and worked to bring my team aboard, to add maximum value to the function, program or business.

---

1. Columbus McKinnon Corporation, When Luck Meets Opportunity (Vince Lombardi), 17 JAN 2011 http://blog.cmworks.com/when-preparation-meets-opportunity/ (Retrieved 21 SEP 2015)

2. Welch, Jack, JACK—*Straight From The Gut* (New York, NY: Warner Brothers Books, 2001), pp. 109-111

3. Dorr, Robert F, Defense Media Network, Gulf War 20th: F-15 Eagles Were the Deadliest Birds of Desert Storm, 07 JAN 2011 http://www.defensemedianetwork.com/stories/f-15-eagles-were-the-deadliest-birds-of-desert-storm/ (Retrieved 05 DEC 2015)

CHAPTER SIX
PEOPLE

The most important single ingredient in the formula of success is knowing how to get along with people.

Teddy Roosevelt
26th President of the United States

As a Business Unit Chief Engineer, one of my most important responsibilities was to attract and select executive leaders and senior managers. In truth, I was pretty good at this—many times other leaders would tap people I selected with their next opportunity. May I say? There was no better compliment than to see my people sought after by others.

Given the complexity of military aircraft, I sought out people with a wide range of engineering skillsets: systems engineering, aerodynamics, structural dynamics and loads, mission avionics and software, flight controls, structures and mechanical systems. I was always looking for folks with experience in key process areas such as the Aircraft Product Life Cycle, and Lean+ implementation.

In addition, I looked for broad expertise with fixed wing, rotorcraft, commercial derivative aircraft, and aircraft simulators. It was important for my teams to have, or be able to tap into, the skillsets and knowledge required to execute current contracts and pursue future business. If any knowledge gaps existed, they were filled with subject matter expertise across the larger enterprise organization. I wanted people who were enthusiastic, self-motivated, and had a positive can-do attitude. They had to be able and or willing to (1) work together with anybody or any team, (2) take on any challenge, (3) demonstrate innovative thinking, (4) obtain results, and (5) adapt to changes in direction or information.

Because so many of these jobs were critical to the success of my team—of the whole company—I often brought in other executives to help review candidate qualifications. Some would be involved in some degree with my programs, others would be "technical" experts, and, of course, someone from Human Resources would always be involved.

Initially, we looked at education and work background with a demonstrated track record of performance, high "leadership" scores on employee surveys, and "diversity."

The interviews were structured . . . at the least, every candidate was asked these questions:

> • Why do you want this job? What you think it is all about?
>
> • How would you set the course or direction for your team?
>
> • In the past, how did you handle a tough problem with a customer and or supplier? Explain the situation, the process you used, and the outcome.
>
> • Provide examples of how you drove results; what

were the situations and explain your role.

● How would you translate strategic direction into actionable objectives and goals? We would include one or two technical questions pertaining to the specific job.

Human Resources would tally up the scores, and then facilitate a discussion of who was the best candidate for the job. The interview team would collectively formulate a recommendation. Next, I would take that recommendation with rationale to the Defense, Space and Security chief Engineer and the Boeing Military Aircraft chief operating officer for approval.

Call it, the mirror image: I often would be asked by folks preparing for an interview—in my organization or anywhere—for my advice. How to prepare? Well, find out as much as you can about the specific job. Learn about the team structure, about the program, about any upcoming milestones, and about the customer. Consider developing a 30-60-90 day plan you would implement if you were selected for the position. Share or leave it with the interview panel as it shows you've been thinking through how you would "hit the ground running."

Valuing Diversity

A diverse team strengthens the collective team and the organization, bringing different perspectives, out-of-box thinking, creativity, and better teamwork. Diverse teams enable the best solutions to organizational and customer needs or problems.

Let me start with a broad definition: "diversity" is much more than race and gender but includes different skillsets, work background, and experience. Diversity encompasses military veterans, people from geographical locations where local cultures might be different, and people who had held leadership positions on international programs. So, how does diversity strengthen teams and improve teamwork? The answer; it enables "diversity of thought" among the team. In order to maximize diversity of thought, it is a leader's responsibility to create the environment where everyone's voice can be heard, even debated, and their ideas are considered for implementation. As one of my colleagues said, "Only with real dialogue, do you gain the advantage of diversity." Generational diversity is a more recent dimension (discussed, below).

I worked to ensure that my teams in both Support and Services and Military Aircraft included all or most of these dimensions of diversity. It was a privilege to lead them both.

Generational Diversity

At the beginning of the 21st century, a fourth generation (Gen Y, the "Millenials") began entering the workplace. They joined Traditionalists (people born by or before 1945), Baby Boomers (people born by or before 1965), and Gen X (people born by or before 1980). Boeing was hiring a lot of people just at this time, and it became apparent that the Gen Y group brought a different approach to work, communication, and interaction with others.

Some Examples of Generational Differences:

Priorities: On "Work Philosophy" or "Work Life Balance," some data suggests that "where their parents lived to work, Generation X works to live, and work-life balance is also a hallmark of this generation." And, "Millennials take the Generation X's work/life balance one step further, to the point where leisure is actually interwoven with work."[1] They are known for their flexibility, and they are often at least initially more comfortable with diversity than other generations.

While the more junior people in the workforce work hard, want challenging work assignments, and many are ambitious, they seemed to do an inherently better job balancing their time at work and time with family and friends. I do think the Boomers can learn a lot from the Gen X's on Work-Life balance.[2]

A 2015 CNN news item highlights a growing change in the workplace—to address the importance of family:

> Adobe announced Monday it is now offering up to 26 weeks paid leave for birth mothers and up to 16 weeks paid parental leave for new fathers who are the primary caregivers. This announcement comes less than a week after Microsoft declared it was offering 20 weeks paid maternity leave on the heels of the Netflix announcement to offer employees—women and men—unlimited paid leave the first year after a child's birth or an adoption.[3]

While it's unclear whether the rest of Corporate America will or won't follow suit on paid maternity/paternity leave, for many people, pressures from family and work are frequently in collision.

Communication. Baby Boomers generally preferred discussing topics or issues face-to-face or by telephone. Generation X prefers email and Generation Y prefers text or Instant Messenger (IM). This was at times a topic of conversation amongst Boomers: "Why do people who sit next to each other communicate via text or IM instead of having a conversation?"

The better to understand some of the new reality, I held round-table discussions with these young Gen Y professionals. To set some context for a productive conversation, I would tee up the conversation with a brief description about myself, my career, and the state of the business at the time.

They were not shy to tell me what they wanted in the workplace:

- Challenging, diverse and, often, "special" assignments

- Participation in Leadership Team meetings, when appropriate

- Increasing opportunity to "learn"—new skills, about opportunities for advancement, about our business in particular and about the industry, in general.

Per their feedback, I would occasionally invite them to my Leadership Team meetings, and give small groups special projects or assignments. The tasks were administrative in nature: analyzing our employee survey feedback, or documenting technical "best practices" for sharing between programs. It was always energizing for me to interact with young professionals. They liked the exposure with senior leadership and they could always be counted on to do a great job.

Toward the end of my career, I had a "reverse mentoring" relationship with a Gen Y engineer. With a lot of hiring over the last 10 years or so in my career, I wanted additional feedback from my mentor on how best to engage and motivate this new generation of talent. I found this to be of great help, an excellent venue to obtain insights and perspectives on workforce questions and concerns.

By 2020, a fifth generation, Gen Z, will be entering the workforce. Be ready, and be early: take the time to engage. Find out what motivates them and determine the best way to leverage their strengths—especially a life-long involvement with computers (literally, beginning before many of them could talk)—to best benefit the business.

Inspiring Performance

Attracting and selecting people for the team was just the first step. The next step was to inspire them to perform, to give their maximum effort, regardless of the task. How did I approach this vital task?

- Provided balanced and actionable feedback on their performance—people need to know if they are meeting, not meeting, or exceeding expectations.

- Set a good example for others. People are watching your feet...so it's important to be a good role model, act with integrity, and do what you said you would do.

- Created a vision for the organization, maintained situational awareness of the business, and communicated with clarity.

- Established a culture that encouraged people to bring bad news forward; problems don't get better with time.

- Rewarded and recognized people for their accomplishments. It is amazing the impact of a simple "thank you" or "job well done."

Evaluating and Developing People

I looked for people—especially, those with leadership potential—who showed me they could:

1. Demonstrate visionary thinking, translate the vision to strategy, then execute.

2. Model technical excellence across their teams or organizations.

3. Monitor leading indicators of performance and head off problems before they become a crisis.

4. Work collaboratively with others and do what's best for customers, the larger enterprise, business unit, programs and products.

5. Raise the level of discussion and performance when participating in strategic planning meetings. Of course, it goes without saying that all were expected to be a person of high integrity and to meet or exceed personal goals and objectives.

As a Business Unit Chief Engineer, I found that the majority of programs chief engineers and direct reports fell into the following categories.

1. Those who delivered results were great to work with and added value to the larger engineering team. They were appropriately rewarded during the performance/salary review cycle and some were placed on executive succession plans. One of my responsibilities was to ensure that we had robust development plans for each to help them in their current assignment and prepare them for their next move. These plans might include special assignments, participation in "Win Strategy Steering

Committees," continuing educational opportunities, including degreed programs and training courses at the Boeing Leadership Center.[4] Also, the Company had a handful of billets for emerging leaders to attend government sponsored training, e.g., Defense Acquisition University (DAU) with senior military officers.

"Delivering results" meant effectively supporting program goals. Further, these men and women established a culture of technical excellence within their own teams, with high levels of first time quality. The work environment and culture were collaborative in nature, and there was a willingness to help other programs with their resources.

Throughout my career, most members of my teams were in this category. Every now and then, performance of one or another might slip, and I would step in to provide balanced feedback and constructive criticism. In my experience, good people appreciated straight talk and would work to improve.

2. Engineering leaders who didn't get results but added value to the organization were generally people who were, simply put, in the wrong job. There were not many people in this category. I would spend time working with them and help them as much as possible. However, usually the action at the end of the day would be to find them a different assignment. For those few leaders, I would communicate to them the value they brought to the team in an effort to not adversely affect their self-esteem.

3. There were, sometimes, a few people with the sort of personality that got in the way of work. Some were

combative with managers, some were intimidating to subordinates, some tried to work around the chain of command to prove they were "superior." What to do? Try to have an open, frank discussion, offer an opportunity for redemption, or . . . well, sometimes the "or" was the only resolution. Speaking of "performance evaluations," usually a number of people have a hand in the process. Thus, a word to the wise employee: develop and maintain solid relationships with everyone, not just your immediate superiors.

Succession Planning

One of the most important jobs of a leader is, insofar as possible, to identify and develop a potential successor. This is especially true with the more successful leaders, because they are the most likely to be moving up, often sooner than anyone would imagine.

However, you should always assume that change will come, in any job, and part of your job is to plan ahead. For example, we had programs that continued for decades: F-15 and F/A-18 have been in production for more than 40 years and the B-52 bomber, while long out of production, has been operational well into the sixth decade . . . with continued support from Boeing. Succession planning provides a seamless transition among managers as people move over or on to new assignments.

Part of my job was to identify possible successors—not just for myself, but for all executives in my organization. Once a year, I collected recommendations from my direct reports, my peers, and senior management. Candidates would be categorized as "Ready

Now" and "Ready with Development." For those identified as not yet ready, we looked for assignments for each that would help them get ready. I can't stress the importance of taking the time to get this right.

There was no more important job as a leader than to develop emerging leaders for the future.■

Chapter Six Leadership Takeaways

• Attracting, selecting, inspiring, developing and evaluating people is a primary leadership responsibility.

• A company's future success is dependent on how well today's leaders prepare emerging leaders. Make it a priority to develop people as part of the succession planning process.

• Teams that value diversity are higher performing teams. Diversity provides out-of-box thinking and innovative solutions to operational and organizational challenges through different perspectives and ways of doing things.

---

1. Notter, Jamie, Generational Diversity in the Workplace, http://www.multicultaraladvantage.com/recruit/group/mature/generational-diversity-in-workplace.asp (Retrieved 05 DEC 2015)

2. Personal Tips on Work-Life Balance:

    a. Take time to reduce stress through whatever health and wellness activities you enjoy. If you don't take care of yourself, you will not be good to anyone. Health and Wellness was a key area of focus for the company for a couple of reasons; (1) the ever-increasing cost of health care and (2) a healthy workforce is a more productive workforce.

b. Make time with your family a priority. When you are home, do the best you can to disconnect from work and be both mentally and physically present.

c. As a leader, encourage people to take their well-deserved vacation. Too many people would say, "I am maxed out" or at the point that they couldn't accrue vacation time. If you are on a trip for most of the week, consider adding an extra day or two to your weekend.

d. It is important to know your people and their priorities outside of work. In today's workforce, many people needed flexibility in their work schedules for all kinds of reasons. Examples included personal reasons, kid's activities, and elderly parent issues. My experience was leaders always did what they could to provide their people with needed flexibility; and the employees found ways to get the job done.

e. When I wasn't traveling, I made it a point to be home at a reasonable time and have dinner with the family. Of course there were exceptions over the years, but very few. If customers, suppliers or teammates were visiting from other sites, we would do lunch...and everyone preferred having their evening free.

3. Wallace, Kelly, "A year of paid parental leave: Vital but how likely?" CNN article, 10 August, 2015 http://www.cnn.com/2015/08/10/health/paid-parental-leave/ (Retrieved 10 Aug 2015)

4. Gunn, Thomas, *Gunnsights*, pp. 29-30, Annapolis, MD: Naval Institute Press, 2007

CHAPTER SEVEN
TEAMS AND TEAM STRUCTURES

A basketball team is like the five fingers on your hand. If you can get them all together, you have a fist. That's how I want you to play.

Mike Krzyzewski
NCAA Basketball Coach

During the development phase of a program, people are generally organized into Integrated Product Teams (IPTs). The "Integrated" infers the various engineering disciplines—structural design, structural analysis or stress, materials and processes, and manufacturing—which are aligned around a specific product like a wing, fuselage or subsystem. In addition, IPTs include functions outside of engineering, such as supplier management, procurement, and finance. "Product Team" simply applies to a part of the airplane, like a wing or fuselage, or a subset of these major airplane components.

Program Organizational Structure (example)

A simplified "program organization" chart shown in Figure 7.1 highlights the key leadership positions assigned to a program and their primary responsibilities.

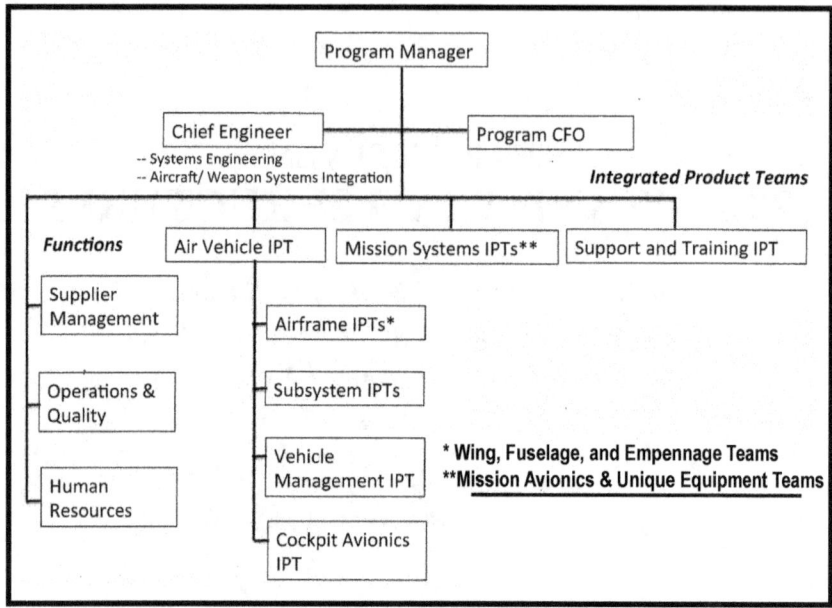

Figure 7.1 Program Organization Chart, Simplified Example

Leadership Positions & Responsibilities

1. **Program Manager:** responsible for executing the program to the contractual commitments, as defined by the customer. This includes responsibility to deliver products through the IPTs, completion of successful program development and demonstration milestones (Chapter Eight), delivery of aircraft, and support and training. The Program Manager is accountable to a Division or Business Unit Level leader.

2. **Chief Engineer:** responsible for technical excellence and technical decision-making on the program. Key focus areas include systems engineering, configuration management, aircraft level and technical integration across the IPTs, integration with other functions, and managing technical risk. This position is accountable to the Program Manager and a Division or Business Unit

level Chief Engineer.

There were a couple of models for IPT reporting relationships, either to the Chief Engineer or Program Manager. My recommendation, during "development and demonstration" through "test and verification," is for the IPTs to report directly to the Program Manager. This model enables the Chief Engineer to more effectively execute his or her responsibilities.

3. **Integrated Product Teams (IPT):** responsible for development and delivering products to the program that meet technical and regulatory requirements, on schedule and cost.

4.; **Functions:** responsible for staffing programs with the right skilled people, and ensure that proven processes and tools are used on the program. In addition, these leaders are responsible for ensuring technical excellence within their domain of expertise on programs. Engineering Functions are accountable to the Chief Engineer on the program. The Functional Leaders depicted in Figure 7.1 are responsible and accountable to the Program Manager.

Integrated Product Teams (IPTs)

The IPTs are responsible to deliver products that meet contractual commitments, on time and budget. IPTs ensure that products are designed for "manufacturing and assembly, supportability, and safety." These teams have full responsibility, authority and accountability (RAA) for a product through its life cycle, to include driving decision making down to the team levels. In a perfect world, these teams have a complete understanding of product requirements and how the product interfaces with the aircraft. Using a nose cone as an example, the key functional and physical interfaces were the forward fuselage and a sensor internal to the nose cone, such as a radar system. All the technical data and information was provided in a timely manner to successfully design, build, test, and support the product. Work scope was aligned with a schedule, budget, and resources to execute the statement of work. Engineers were provided with proven processes and tools to do their jobs.

> *Technical lead engineers were in place to advise, assist and check work—I can't emphasize the importance of these people and positions to help ensure first time quality engineering.*

Integrated Product Teams are an efficient and effective structure, especially in the development portion of a program. It is imperative, though, that experienced engineers are engaged in key technical decisions and recognize that it is their responsibility to elevate discussion on technical risks, issues, and help needed. Teams were empowered to make adjustments to their designs, plan and schedules as long as the end goal of "released engineering" complied with all requirements, completed on time and within budget.

Team performance is measured, and team-members are encouraged to ask for help when needed, not to try to gut it out. One of the first programs to fully embrace the IPT concept was the F/A-18 E/F *Super Hornet*, which applied this team approach through years of strong and effective leadership.

Once development was completed and the program transitioned to manufacturing, engineering staffing levels dropped significantly and skill mix requirements changed. Added to the team:

experts in "liaison engineering" to interpret engineering drawings or work instructions, and to help resolve manufacturing issues. Further, depending on number of programs in production at a given site, Integrated Product Teams provide manufacturing support on several programs.

Shared Resource Teams

An excellent example of this approach was a structure of IPTs put in place in the mid-1990s to support three production aircraft programs, the F-15 *Eagle*, AV-8B *Harrier II* and T-45 *Goshawk*. This was a dual-hatted responsibility for me when I was the F-15 chief program engineer.

There were four IPTs within the shared resource team. They were (1) airframe and subsystems (electrical, fuels, hydraulics, environmental control, etc.), (2) avionics, software, integration and test, (3) systems engineering, including configuration and product data management and (4) test and evaluation.

One advantage of IPTs that support multiple programs is reduced cost, by leveraging critical mass of some of the engineering skillsets. It also enables knowledge transfer between the more senior engineers with newer college hires. Where we had specific development efforts, we did put program specific IPTs in place, mostly in mission systems and software to achieve the level of integration required to complete the statement of work. On the flip side, program managers don't always like their program tasks taking a backseat to other programs that are getting higher priority. The effectiveness of a Production Program IPT structure was only as good as the people leading the team, understanding and managing priorities, and communicating with internal customers. Your decision as a leader in evaluating this type of structure is around trading control of resources for cost and what is best for your program or business.

Engineering Centers of Expertise

While many engineers are working on development, production and support programs, not everyone is aligned to an IPT. For specialty engineering skills—for example, nonlinear finite element modeling, fatigue and fracture mechanics, or flight control law development—we generally "cored-up" people to Engineering Centers of Expertise. Simply put, there were generally not enough people with these specialized skills to populate program specific IPTs. It was more efficient for teams to come to the Engineering Center to buy specific services.

Our Structures Engineering Center of Expertise in Philadelphia was a great example of a highly performing team. Some examples:

1. Successfully completed a sizeable work package for the development of a Wing Aerial Refueling Pod (WARP) pylon for a tanker program

2. Resolved a significant technical problem, with an "as-built" nonconforming condition on an in-production fighter aircraft's empennage, to preclude affecting the delivery of jets and spare parts.

3. Created a finite element model that represented an out-of-production legacy platform for the purpose of evaluating life extension

All three projects were executed by a team which had strong technical knowledge, highly proficient in the necessary tools, and had a disciplined process. Such as? Understanding the statement of work, asking questions to obtain clarity, and utilizing peer reviews to ensure first time quality. Well. . .those are technical elements—certainly important—but in my judgment, the team stands apart because of their eagerness to help others, to communicate frequently with internal customers, and they delivered results. They exemplified the definition of teamwork, and for me, it was always a pleasure to work with them.■

Chapter Seven Leadership Takeaways

● Establish team structures, such as Integrated Product Teams, with clear responsibility, authority and accountability. Push technical decision-making down to the lowest level to increase team efficiency, to enhance the ability to adapt to changes in direction or to face problems.

● Ensure that technical lead engineers are in place on Integrated Product Teams to advise, assist and check work, and to ensure technical risks, issues, and the need for "help" are elevated up the management chain, when appropriate.

● "Coring up" specialty engineering expertise can be a highly effective organizational model for an organization at a division or business unit level.

CHAPTER EIGHT
AIRCRAFT PROGRAM DEVELOPMENT AND DEMONSTRATION

Our armament must be adequate to the needs, but our faith is not primarily in these machines of defense but in ourselves.

Fleet Admiral Chester W. Nimitz

There are many dimensions to military aircraft programs. As in any business, it's important to know the customer-assigned missions, current resources and capabilities, future requirements, and the environment in which they operate. In this context, "environment" does not mean the weather, but the culture . . . personnel policies and training, rules of engagement.

The acquisition of military systems is complex, systematic, and time-consuming, involving a mix of planners, users, designers, contractors, testers and, always, whatever agency is providing the funding. It is at times a delicate balancing act between requirements, cost, and schedule, always driven by a need to deliver a product on-time, on cost, and mission-ready.

The Systems Engineering process, as depicted in Figure 8.1 and described in detail, following, is the framework for managing and integrating development of military aircraft through full life cycle. First, the customer identifies a need or want, and at the end, the product has been fielded and is operational. This disciplined process is proven to result in an aircraft weapons system that meets the requirements of the customer with first time quality, on time and within budget. In addition, the process provides a mutual understanding of the health of the program to both the customer and the industry team—the prime contractor, primary suppliers and sub-tier suppliers.

First time quality in engineering doesn't happen unless the organization has engineers with the domain knowledge and expertise consistent with all phases of the product life cycle.

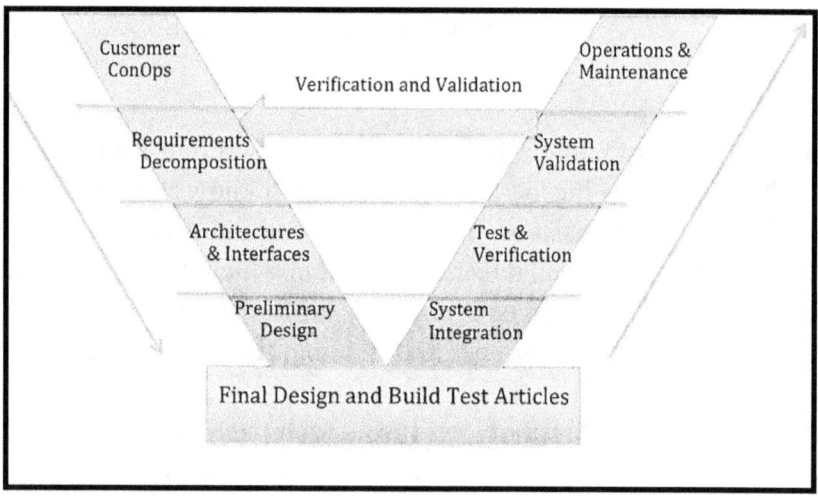

Figure 8.1 Systems Engineering "V" Chart.[1]

THE PRODUCT LIFE CYCLE

The process begins when a customer (typically, the Department of Defense) develops a **Concept of Operations** (ConOps)—

defined as "a verbal or graphic statement that clearly and concisely expresses what the joint force commander intends to accomplish and how it will be done using available resources."[2]

A ConOps might include requirements for interoperability with other weapon systems—in the air, at sea, and on the ground. How, for example, will the aircraft communicate and work with each other while engaged in training, air combat, or when ground troops are calling for air support. Once the ConOps is accepted, a new program is identified and a Request for Proposal (RFP) is issued to solicit bids from military contractors. The RFP defines the requirements; here, a paraphrase of requirements for a Strike Aircraft (e.g., F-15E, F/A-18):

- Mission readiness requirements.

- Primary mission capability: Air-to-Air, Air-to-Ground.

- Speed-altitude flight envelope, maximums and minimums.

- Maneuvering load factors for specific aircraft configurations and status (such as, fuel state and weapons loading).

- Aircraft range, ability to loiter (that is, remain on station) for specified times.

- Environmental factors to support operations in austere locations.

- Two seat tandem cockpit.

- Digital mission unique avionics and mission equipment.

- Ability to target threats and launch (defined) weapons at specified distances.

- Aerial refueling capability.

- Logistics support and training.

- Development and Demonstration timelines.

The various manufacturers who receive the RFP (who likely will have known a lot about this program for some time) will assess and, if interested, launch a careful and thorough evaluation and response. This would define the offering, explain how it would comply with the customer's most important requirements, and price out the statement of work.

There was much more, of course, to the process, most of it long before the response would be sent to the "customer." To begin with, the RFP was not an isolated hands-off solicitation, but the beginning of a running dialogue to explore ambiguities, ensure clarity and enhance understanding.

From my "engineering" perspective, we must: ensure that our response complies with the customer's most important and technical requirements. We had to understand each, and describe how the proposed solution would meet each, with specificity and clarity. We would highlight the processes we would use to design, build, test, and certify the product, and outline our organizational structure, including suppliers involved, and explain how we would measure program performance. Highlighting successes in developing and integrating similar and complex capabilities on previous programs was always a plus, which generally suggested reduced risk during development.

How would we staff the project? Clearly, you can't just jump into a mega-multi-million or billion-dollar contract without a great deal of advance planning. We had to ensure the right skills would be in place at the right time throughout the projected schedule. The functional team defined the sources of staffing—such as, de-staffing other programs, hiring new college graduates, recruiting at job fairs, and encouraging job rotations to obtain the right balance of skills and experience.

We would issue our own version of an RFP to potential suppliers, for key items that traditionally are procured from outside—for example, mission avionics and software—with consideration for level of development required and associated technical risks.

At about the same time—well, everything was going on at about the same time—we would assess the "likelihood and consequences of technical risks." For example, if relatively new technology was to be employed, what was the track-record to date? Was it mature enough to give confidence? We would have some back-up plan, a work-around or an alternative technology. Alternative? Possibly, similar equipment already developed for other Boeing programs . . . if it would meet technical, contractual and regulatory requirements. One of my particular responsibilities was to attest that our response was a sound technical approach that would meet the customer's most important requirements, and that we could, indeed, to the best of my knowledge and belief, execute the contract on schedule and on cost.

I should note, there was another component when bidding for international sales: it was called, "Offset," contracting for "direct or indirect" work with some local, in-country engineering and manufacturing companies. "Direct" meant the program would put a predetermined percentage of work on this contract in a local industry. "Indirect" meant we would assign a predetermined percentage of work from other programs to local industry. Whether direct or indirect, the work content that would be performed in country would have to align with the skills and capabilities of local industry.

For a thorough look at the RFP/Response process, see the memoir by former McDonnell Douglas executive for Sales and Marketing, Tom Gunn.[3]

Winning New Business

Vince Lombardi had another great saying, "Winning isn't everything, it's the only thing." But … what does it take to win a competitive defense contract? Let me raise the curtain and look inside.

It's important to understand that the United States government's acquisition process has significant depth, breadth and complexity. There is a Defense Acquisition University (DAU), where military and government civilian personnel attend courses that are weeks and months long to obtain the necessary education and training required to be proficient in the acquisition process. Contractors are invited to send a few people to attend these courses, which provide our people with not just deeper knowledge of the acquisition processes, but the opportunity to build new and strengthen existing relationships with customers.

The current environment that customers evaluate the contractor's RFP responses and select the winner, based on my recent personal experience, is based on the response being (1) technically acceptable and (2) having the lowest price.

Technically Acceptable. I have outlined the mechanics and design consideration highlights, so to speak, of responding to a RFP (see the previous half dozen paragraphs). Bottom line: the technical offering had to address the customer's most important requirements, affirm that the program was executable, and that the right people would be in the right team structures, and processes and tools would be in place to enable the highest level of productivity. In addition, we affirm that the requirements are well understood, and design and integration maturity is at a level that the risk of implementation can be managed to support development and demonstration timelines. The Case Studies in Chapter Ten on the F-15SA (Saudi) variant, T-38C avionics upgrade, A-10 re-wing and U.K. *Chinook* program bring these mechanics and design considerations to life.

In addition, keys to crafting a winning 'technical section' proposal response requires:

- Knowledge of business development processes.

- Feedback from sources within and outside the company with expertise that could provide insight and perspectives on what was most important to the customer.

- Exceptional communication skills for oral presentations to the customer.

- Technical expertise in operational innovation and competitive differentiation to include rapid prototyping for product demonstration.

- Engineering leadership commensurate with the RFP request, capable of building and leading the right technical team with domain knowledge and expertise.

Winning Price. Business unit (or higher) and business development leadership would derive what they collectively felt was the winning price, based on analysis by internal and external organizations. In addition, there was consideration of business strategy such as gaining access to new markets, and cost to execute the technical proposal section.

When evaluating the contractor's RFP response, customers don't just look at the contractor's price at face value. The acquisition customer considers a contractor's basis of estimates, including pricing assumptions backed with facts and data. Past performance, or having a track record on previous development programs or R&D projects, is a key consideration in evaluating a contractor's ability to perform to their proposed plan. Design complexity, or—more important—design simplicity, combined with mature technologies, are key factors in ensuring an executable program with a manageable level of risk. Considerations also include the industry team, meaning teaming partners and supplier; specifically, do they have previous experience in working together? And, if so, what was the collective past performance? All together, this is why successfully executing current contracts is so important toward winning future business.

It is not uncommon for customers to evaluate a contractor's proposal, and based the customer's experience with other similar types of programs, to add cost factors to a contractor's price. This is called an estimate of "price realism". The price used for evaluation and selection is based on the government's assessment of price.

Post Contract Award
Once the customer provides a contract and statement of work,

we would begin Engineering, Manufacturing and Development (EMD). The contract would define "capabilities" (performance, payloads, weapons, sensors, defensive systems and other mission critical equipment). The contract would also include Key Performance Parameters (KPP), which were quantitative metrics reviewed at a specific frequency by the customer. The T-38C Case Study in Chapter Ten highlights an example of KPP.

Working down the left hand side of the Systems Engineering "V," **Requirements Decomposition** is a process that takes the customer's requirements and flows them through a Work Breakdown Structure (WBS).[4] Requirements consist of (1) contractual requirements, (2) regulatory requirements and (3) derived requirements. Derived requirements are contractual or regulatory requirements, not specifically stated in a government contract or regulatory manual, which are defined by systems engineers in a way that the aforementioned requirements can be met by IPTs. Derived requirements are flowed down to the product teams during development.

Stable requirements are key to achieving first time engineering quality. Each team has a defined statement of work or scope, a set of requirements, a budget and a schedule. A contractual requirement might be "the aircraft shall be able to pull 7gs of load factor through the operational envelope (e.g., Mach-altitude and specific aircraft weights.)" In order to satisfy this one requirement, multiple working teams would coordinate the following steps:

- Planning, conducting and completing wind tunnel testing (handled by Aerodynamics, Loads, and Test teams), which includes model fabrication, instrumentation and checkout.

• Reducing data from the wind tunnel for input of aerodynamic coefficients into a database combined with mass property inertia data (Aerodynamics, Loads, and Mass Properties or Weights teams)

• Using a Loads Maneuvering program to survey for the symmetrical and asymmetrical design load conditions throughout the operational flight envelope. An example of a symmetrical maneuver would be a pull-up with no roll or yaw rates or accelerations. An example of an asymmetrical maneuver would be a rolling pull-out, which would combine a pitch, roll and yaw that generated some level of sideslip. (Loads Team)

• The definition and documentation of all critical load conditions for every part of the aircraft structure. (Loads Team).

This was just one of many (hundreds to thousands of) specific requirements, to be passed down to multiple teams through the WBS. "First time quality of loads" for the aircraft structure was essential; "first time quality" means from the very beginning, (not, oh, "I guess we missed something, but don't worry, we'll fix it"). "Loads" translate directly to composition and sizing (material, depth, thickness) of structural members (spars, ribs, frames, etc.) Once the structure is sized, electrical and mechanical systems engineers begin routing wire bundles, fuel lines, etc. If the loads are not done right the first time, there will be significant rework not just for the structural engineers but for the systems team which now has to re-define the routing of bundles, fuel systems, etc. Ensuring accurate loads, and integrating both timing and sequencing of structural sizing with subsystem definition is a prime example of the role of a program chief engineer with the IPTs.

"Loads" played a role not only in "structures," but in almost every other component. One example: avionics hardware had to withstand the load factors, but, to emphasize the complexity of this whole process, the hardware had to handle environmental factors such as vibration and temperature which might be in play when there are no loads, even before the aircraft leaves the ground.

Next step: defining Architectures and Interfaces

"Trade studies" (that is, a "trade-off" between options) are explored to find the best solution that meets the customer's requirements. For example, we might bring in engineers with expertise in configuration definition, structures, flight controls, systems (fuels, electrical, hydraulics) and mission avionics. They would lead architectural trade studies that define key physical and functional interfaces, which would meet the mission scenarios during flight test and fleet or operational usage. Then, software systems engineers would develop requirements for software design.

Using a simplified fuel system as an example, a trade study would define fuel line routings and fuel system attachments to wing and/ or fuselage structure and placement of valves and pumps (which would regulate fuel system pressures for different aircraft speeds, altitudes and maneuvers). How would fuel flow and fuel rate be managed to accommodate changing aircraft weight, balance, and fuel quantity? Also, what would happen if a component failed? How to ensure system safety?

I said, above, "a simplified fuel system" but you may begin to realize that there is nothing "simple" in any of this . . . and this stage of the process is before preliminary design is even started. Add trade studies to define the airframe structure, hydraulic and electrical systems, mission avionics, flight and controls, and you begin to get the idea . . .

However, before turning all of this over to the designers, the company would hold a "Gate Review" focused on the quality and completeness of the engineering and related trade studies (e.g., Supplier Management and Procurement, Operations and Quality). A "Gate Review" brings a team of independent subject matter experts, not assigned to the program, to ensure that the appropriate level of design maturity has been met. (See Chapter Eleven.)

At this point, we will have created the "aircraft configuration" and were ready for a major program milestone, the Systems Requirement Review (SRR) to be led by the customer.

Next, after the Systems Requirements Review (SRR) affirms all factors, preliminary design begins. This is a critical juncture (re-read, the half dozen paragraphs above). From this point, changes to requirements often drive rework, which may affect productivity, performance, and milestones. Trust me, you don't want to be there.

Now, a caution: requirements are not always fixed and immutable. Changes in mission profile, changes in the threat, advances in new—or failures revealed in old—technology, each or all may have an impact, and must be considered. If so, a revised statement of work is put before a Change Board, usually chaired by the program chief engineer or program manager. The objective: yes, to get agreement that the change would be beneficial but also, to ensure budget and schedule relief from the customer.

Preliminary Design—At this phase of the program, teams and suppliers begin the design process. SRR has been completed and action items closed. Design and integration of new technologies should be at a defined maturity level. Conceptual layouts show details of structures, specifically at key interfaces. Mechanical and electrical systems have been designed to the next level of maturity, consistent with the architectures and interfaces defined in this phase. ("Next level of maturity" would include fuel, hydraulic and electrical systems running through the aircraft, to include where tubing and wire harnesses tie to structure.)

Loads were released for structural sizing, with conservative assumptions applied that were consistent with company and industry best practices for areas of local load introduction. Materials have been selected and characterizing material allowables were being worked. Keys to first time quality were completion of external loads and material selections with design allowables in advance of final design.

Cockpit and mission avionics preliminary design would be underway, including software. Aircraft level integration analysis was also underway. This included calculated design weights from preliminary designs assessed to ensure that KPP's such as aircraft range and other performance measures showed sufficient margins. At this point in the product life cycle, the objective is to mature the areas of highest technical risk and identify/resolve any showstoppers. Teams are reviewing designs with suppliers. The interdependency between teams associated with technical integration and coordination continues to increase quickly. Structural engineers are creating conceptual structural layouts, and need feedback from systems engineers. Structural engineers have to work with flight control engineers to ensure that external loads during maneuvers will be consistent with the aircraft's fatigue spectrum. Systems engineers ensure that functional design of the fuel system can't generate pressures that exceed the design capability analyzed by structural engineers.

As with the SRR, a team of functional or independent subject matter experts not assigned to the program work closely with the program teams. They ensure the appropriate level of design maturity, through a comprehensive review of artifacts. The closure of this phase culminates in a customer Preliminary Design Review (PDR).

As an instructive footnote, of sorts, at this transition between Preliminary Design and Final Design, let me review the evolution of product definition processes and tools since the late 1960s, early 1970s. Back in the days when engineering drawings were hand drawn with ink on mylar. Today, computer aided technology is a key enabler to engineering . . . everywhere.

Design for Manufacturing and Assembly (DFMA) would be the focus of electronic drawings and electronic engineering through the 1990s, with the inclusion of manufacturing engineers and tool designers embedded into IPTs. Their inclusion provided real-time feedback on designs, to enable the factory-floor build process. As part of the engineering drawing release process, and through inclusion of these engineering disciplines into the IPTs, manufacturing work instructions, tooling fabrication and assembly work orders were released concurrently. While this approach shifted engineering staffing levels higher to the left, it reduced the hours at the tail end of the design process. Further, it would flatten the learning curve and enable production to build the first unit at a reduced cost from "business as usual." The other benefits to DFMA on legacy aircraft was that teams would combine multiple "sheet metal" parts and convert them into one piece machined parts. Further, this had a compound effect on reducing fasteners and fastener types.

DFMA evolved into the first decade of the 2000s and beyond with innovations by tool design and manufacturing engineering functions working together with the IPTs. One approach was the use of major structural parts such as bulkheads, frames or formers, and spars as "tools," instead of the traditional assembly jigs (AJs). These structural parts located mating parts in a way that enabled better fit and reduced variation in the assembly process. The results of implementing this approach were reduced cost and cycle time, and improved quality.

"Design for supportability" and "design for ergonomics or safety" would become key considerations; for example, mechanics and electricians could access the product easier and in a way which reduces or eliminates factors that cause pain or discomfort. The addition of operations, quality, safety, and logistics expertise into the design process provided significant value added to new product designs or redesigns of legacy aircraft. Specifically, these organizational functions were given "a seat at the table" to participate and influence design decisions.

"Immersive collaboration" was a follow-on evolution of product development processes and tools. In this, all stakeholders, whether it be developing a new aircraft, designing improvements to legacy aircraft, or even looking at a real-time assembly problem on a production line, have the capability to meet in a Virtual Reality (VR) environment. This consists of a meeting room with computers, projectors, screens and people working together, viewing the same data from different perspectives. In addition, immersive collaboration, using VR, is enabled by real time engagement across sites where design and/or build teams—at different geographical locations, through information technology networks and telecom technologies—can participate as if they were present in the same room.

The collective team can also view 3D data or stream live images from a production line. The objective was for all stakeholders to immerse themselves, leveraging the environment and data/information to converge on optimum solutions. This meant ensuring that products met customer requirements, defining how the build team could best assemble the weapon system, and easing maintenance for the warfighter.

As a real example, while in BMA, a program team on a fighter aircraft program leveraged immersive collaboration on a redesign of a sizable fuselage component. The team was realizing significant improvement in first-time build quality and factory floor productivity at the time of my retirement. The A-10 Case Study (Chapter Ten) reflects these design considerations and processes.

Final Design and Start Build of Test Aircraft. Preliminary Design Review (PDR) has been completed and action items closed. Final external loads and material allowables have been defined and released to all teams. Materials such as aluminum, composite materials, even titanium, had been considered and selected for structural analysis, given the mission environment. Size, weight and power requirements for cockpit and mission avionics and other components had been designed and integrated into the aircraft. Design of electrical wire bundles (documenting routing and installation throughout the aircraft) had been completed, as were fuel, hydraulic, and environmental control systems.

At that stage of a program, engineering staffing for all technical disciplines was at its peak. Engineers from all disciplines were assigned to teams to mature the design details, to include completing analysis and documentation. Engineers from Centers of Expertise were on board helping the team with specialty skills, as appropriate. Technical lead engineers were assigned to advise, assist and check work of more junior engineers. Physical and functional interfaces had been defined to include back-up structure at or behind local load introduction points such as a wing to fuselage major join or wing to pylon connections.

The end of final design should complete approximately 90 percent of drawings or models. This includes structural and system analysis nearly complete by the Critical Design Review (CDR) milestone. System safety analysis, which assumed equipment failures or failure scenarios, had been considered in the design. Design for manufacturing and supportability had been incorporated. Software designs that include functionality such as display formats, cautions and warnings, had been completed. Coding of software was well underway. Supplier designs, which generally include mission avionics, associated software, and other mission equipment, had been nearly completed. After subject matter experts completed the Gate Review, this phase culminated in the customer CDR.

I can't over-emphasize the need to achieve the right level of design maturity and definition of the technical baseline at each milestone defined above. If this doesn't happen, a program or project can and will travel technical risk from one product life cycle milestone to another. In other words, the scenario is set for significant rework, impacting cost, schedule, and commitments to the customer—which might require root cause analysis, new design and implementation to create a fix, followed by incorporation in the test aircraft.

So, how does one know if you have the right level of maturity and definition of the technical baseline? If the requirement is to have a 90 percent completed drawing release, don't just look at the performance-to-metrics in total across the program, you must review performance-to-metrics for each IPT. Consider whether a team or teams are significantly behind and don't have a lot of drawings to affect the aggregate data at the program level. Also, check to see that all analysis is completed and documented. Those could be red flags.

- Ensure all Supplier CDRs are completed in advance of the Program CDR. If not, it is highly likely, based on personal experience, there could and would be changes that affect the design after CDR.

- Determine if sufficient margin on KPP's (such as aircraft weights, range and other mission critical measures) exist.

- Conduct Virtual Reality design reviews of physical installations, with focus on key interfaces.

- Ensure Peer Reviews have taken place on designs and analysis.

To be a contender in a highly competitive market, you must demonstrate your ability to deliver first-time quality engineering and the required level of maturity for each product life cycle phase, as defined by Figure 8.1.

By this point, builds of flight test aircraft and full-scale static and fatigue test airframes were underway, which included hardware being built for assembly. Engineering had been spending time supporting the factory floor, addressing questions and resolving discrepancies with the released drawings. Engineers across the IPTs and suppliers were writing test plans for integration and test. Laboratories should be in place and ready to verify that products—components, systems—were working as designed.

Supplier Technical Oversight. While this topic isn't one of the Product Life Cycle phases, it is a critically important process through every phase of development. Suppliers are partners and considered part of the IPTs. Their success is directly related to the success of a team and program. On military aircraft programs, suppliers can represent 50 to 70 percent of the total program cost. In many cases—for example, jet engines—products or systems were competed separately by the government. Cost of mission unique avionics such as radars, defensive systems, displays, and software were the higher cost development items. Structural parts, mechanical systems such as auxiliary power units, raw materials, and fasteners were examples of components supplied as either development items or commodities.

The most important step in working with a supplier is to provide a contract and statement of work with clarity. Integrated Product Teams, through the Systems Engineering or Product Life Cycle process, flowed down applicable requirements and technical data in both an Interface Control Drawing (ICD) and Source Control Drawing (SCD) to suppliers who then designed their parts of the system. Using the fuel system example, an SCD would contain design fuel pressures combined with aircraft maneuvering load conditions and environment conditions, such as temperature extremes for structural and functional analysis.

Here is a representative process for engineers assigned to IPTs with responsibility for supplier technical oversight:

1. Provide an approved statement of work, prepared by engineer(s) on the responsible IPT prior to contract award.

2. Review the supplier's plan to accomplish the

statement of work; make sure it was integrated with the program schedule and resourced appropriately with the right skills at the right time.

3. Ensure that requirements and key interfaces are understood, especially unique company requirements, such as "failsafe" design. Boeing requires, for example, that if one of several wing-to-pylon interface attachments fails, the remaining physical attachments would be able to carry limit load.

4. Ensure that any product being reused, or which had been developed for another contract, meets technical and regulatory requirements.

5. Understand supplier development engineering processes and ensure they are being followed.

6. Keep tasks—such as providing loads and material allowables—off the critical path.

7. Review performance-to-plan weekly, and provide key onsite expertise to continually strengthen the relationship.

8. Use Technical Interchange Meetings (TIMs) between prime contractor and supplier(s) throughout all phases of the program. Document meeting minutes, to preclude future disagreements on technical items.

9. Use a venue of multiple "side-by-side" technical reviews between the engineers assigned to the IPT and suppliers to review design details and analyize approaches. Recommended actions: (1) Invest the time necessary to go through the details prior to engineering release of drawings. (2) Review and concur with key technical assumptions during the preliminary and final

design phases. (3) Ensure technical assumptions are anchored with appropriate analysis and or test data.

10. Ensure maturity of technical baselines at key program milestones. Failure of this step can result in traveling risk to a downstream phase of a program.

11. Participate in supplier's gated milestones such as preliminary design, final design, and test. Ensure artifacts are reviewed and complete in advance of program's gated milestones.

12. Review and approve final analysis and artifacts required for airworthiness certification.

If suppliers didn't perform to plan, it was likely because not all of the process steps above were followed. If and when suppliers get off track, increased team intervention should generally kick-in. Depending on the severity and criticality of the problem, higher levels of leadership from both companies should engage to ensure prioritization and determine "help needed" (to include the resources to bring technical issues to closure). The most important advice I can offer: follow the 12 steps. Err on the side of too much communication, or too much collaborative engagement. If you skip steps in the STO process, you run an increasing risk of not getting the desired product when expected. Employing a "trust but verify" approach is highly recommended. (Chapter Eleven).

Both the contractor and supplier have a responsibility to deliver compliant products to the program. The most important goal of any team is to do what is best for the program; in our business some used to say, "The airplane's the boss." During times of contention, my approach for diffusing conflict was to communicate, "We have to do what's best for the airplane and the customer."

Think about how much technical executive leadership must become engaged once a problem becomes an issue. If that level of talent had been engaged up front to ensure the team started right, their chances of finishing right would be greatly improved.

The System Integration Phase. A Test Readiness Review (TRR) using documented criteria must be conducted for entrance into Integration. This is the phase of the program where software is brought together in System Integration Labs and tested to ensure that all worked as intended. Next, the hardware / software interfaces would be verified; any discrepancies would be noted and referred for correction. For flight control testing, hydraulics are connected to manned flight simulators to portray the aircraft as realistically as possible prior to flight. Some programs might have a 'Wet Lab' to test critical functions such as fuel flow, or to ensure that mechanical systems operate as planned. This is also the phase where mission avionics and equipment are put through rigorous environmental qualifications—vibration, temperature extremes, endurance and other conditions. At a minimum, mission avionics and equipment must complete a "safety of flight" qualification, or some subset of full qualification, in advance of flight test.

In my opinion, this was and is a critical phase of the program. Why? Because this is the point at which you are starting to build, integrate, and test structure and systems. You expect to find anomalies and/or problems in this phase and must be prepared to correct them. Hundreds and thousands of tests are conducted in every program's collective phases of development.

Newly developed aircraft, or aircraft with major structural changes, generally go through full scale static and fatigue testing—where an airframe is built, then secured to a strengthened floor in a test facility where the design load conditions are applied. Strain gages and other instrumentation, such as accelerometers, capture test data that is correlated with analysis during ground testing and development flight-testing.

A second aircraft is put through tests which represent at least two lifetimes of flying—sometimes extended to three lifetimes, in anticipation of a long service life. Fatigue testing can take many months, even years, based on customer requirements. On new aircraft programs, fatigue testing must be ahead of flight-testing by some margin.

Before first flight, ground vibration testing is conducted to verify predicted aeroelastic and aeroservoelastic behavior. Aeroelastic behavior is the interaction of aerodynamic forces, mass properties or structural weights, and the structural stiffness when exposed to fluid flow such as gusting air loads. Engineers consider this interaction in the design of aircraft such that in turbulent conditions, structural components such as a wing will damp out after being impacted by wing gusts. The data is used to validate predicted dynamic responses for phenomenon in flight, such as flutter and buffet. Static and fatigue testing data is correlated to strain levels predicted by modeling and analysis. Ground testing also includes electromagnetic compatibility and interference testing.

Problems encountered during System Integration are corrected, prior to full-scale ground and flight test. In the T-38C Case Study, you will see an example of a program that didn't make use of a hardware-software integration lab, where I describe the associated program impact. Depending on findings from System Integration testing, such as the F-15 Saudi (SA) program, flight testing may begin by providing incremental capability. Given there is always a Safety-of-Flight Review Board in advance of first flight and some follow-on flights, there are accepted practices to use Temporary Operating Limits (TOL) to start flight testing while findings from System Integration are still being addressed. Programs generally do not bid testing incremental capability, so they realize additional cost and schedule impact to update and qualify additional software builds and correct design deficiencies.

Test & Verification. The systems engineers in the Requirements and Definition stage, who had the responsibility to decompose, derive and flowdown requirements to the product teams, had the responsibility for requirement verification at the system level. This stage of the Product Life Cycle starts with verifying requirements at the subsystem levels, such as Level 5, then Level 4, etc.[4] Eventually, requirements verification is rolled up to the system level or Level 1. Verification at the subsystem level is generally accomplished by analysis, inspection, or test. System Level verification is predominately verified by ground and flight-testing. Once ground testing is completed, flight testing exercises the aircraft through a lengthy program, flying within and to the extreme corners of the operational flight envelope, to check for flutter and handling qualities.

"Test and airworthiness certification" is a rigorous process that ensures the integrity of the aircraft. For the Test and Verification Phase, aircraft are exposed to the complete operational envelope. Test pilots check the aircraft against design requirements, ensuring the weapon system functions "as-designed" and meets performance objectives. During Test and Verification, the program enters Development Testing, a phase where a combined government-contractor team flies the aircraft against contractual requirements. Feedback from the testers, combined with data from the instrumentation, enables a comprehensive evaluation.

Engineers use the data to compare measured external loads against predicted data (which was based on wind tunnel data and the output of the Loads Maneuvering program). This verifies that the measured data matches the predicted data. If so, it is safe to fly across the operational envelope, demonstrating that the aircraft would accomplish mission objectives. If measured and predicted data don't match, it is not exactly "back to the drawing board" but it is likely that restrictions will result in a temporary operating limit until the data is better understood.

During this phase of the Product Life Cycle, we conduct a Production Readiness Review (PRR) with the customer, incorporating rigorous, documented criteria, which—when successfully completed—should lead to a decision for Low Rate Initial Production (LRIP), specifically the build of production aircraft for Initial Operational Test and Evaluation (IOT&E) and deployment of the aircraft to the customer. This acquisition strategy requires validation from the PRR that the resources, supply chain, and both manufacturing and quality processes are stable and in place to produce aircraft at a specified production rate.

Systems Validation. Once Development Test is completed, production representative aircraft (but generally not fully instrumented) are provided to flight crew and maintainer operational testers. In IOT&E, the aircraft are flown and maintained as they would be when part of an operational squadron or air wing. The aircraft flight and maintenance manuals are provided to the aircrew and maintainers. When the IOT&E is successfully completed, the aircraft weapon system is declared "operationally effective and suitable," and therefore ready for training, fleet operations, and combat. Successful completion of IOT&E is required for a Low Rate or Full Rate Production decision. In one case study in Chapter Ten, the aircraft was deemed "operationally effective and potentially suitable," which meant there was additional work to be done after a Low or Full Rate Production decision.

Operations and Maintenance. By this point, operational testing and evaluation has been completed and the aircraft weapon system has been found to be effective and suitable. Warfighters are flying initial production aircraft, A production decision has been made and, depending on contract terms, funding for a block, or specific quantity of aircraft already predetermined, has been funded for manufacturing—until the planned quantity of aircraft and spare parts are built. Some programs are awarded a multi-year contract, which enables both Boeing and its suppliers to plan longer production runs and take advantage of lower costs associated with economies of scale. After-market support is described in detail in the next chapter, to include how Boeing works together with the warfighters to help them meet their readiness objectives.

Engineering Support of Manufacturing

During Engineering, Manufacturing and Development (EMD), test articles and flight test aircraft are built. This is the first time the engineering drawings and process specifications, work instructions, and tooling are used by the build or factory floor team in fabrication and assembly. Engineers familiar with the design work side-by-side with mechanics and electricians during build of these aircraft. This is a proven practice to ensure that the above-mentioned engineering media can be used to build the aircraft and meet initial cost and schedule targets.

Engineers address questions and help interpret drawings, effect the right changes such as re-sequencing tasks to improve build efficiency, or correct part definition, work instructions, or even tool designs. It's also an opportunity to ensure that critical or key characteristics that are documented on the drawings can be achieved. When new materials are being introduced, engineers can plan to spend time at either a supplier facility or in fabrication shops to ensure the defined processes and tools can build parts that meet form, fit and function requirements.

After an LRIP or Full Rate production decision is made, the IPT's remain engaged to ensure that production learning-curve targets are achieved and key processes remain stable. As during development, there are likely to be requests for changes, either to correct an error in the engineering or bring forward a product or process improvement. Program change boards are convened to assess and approve such changes; since it is likely that production is underway, and unless these are "flight critical," they will be introduced with an aircraft not yet in production (identified by the "tail number" it will carry from birth to retirement). Another key task of the IPT's during production, as described in the *Harrier* wing example, is to disposition non-conforming material in conjunction with authorized liaison or Material Review Board engineers assigned to the factory floor.■

Chapter Eight Technical Takeaways

●The Product Life Cycle provides a systematic approach to execute a program with disciplined processes. Strong systems engineers and robust systems engineering processes are keys to first time quality. You can't "inspect-in" engineering quality, it has to be designed-in right, the first time.

● If you rely on suppliers for engineering, institute a systematic process to perform supplier technical oversight. This includes a contract with a clear statement of work and engagement throughout the effort, to ensure the product meets the technical and contractual requirements.

---

1. Systems Engineering Fundamentals, Department of Defense—Systems Management College, January 2001. http://www.dau.mil/publications/publicationsdocs/sefguide%2001-01.pdf (Figure 7.1, p.65) Retrieved 31 January 2016). Note: The author used this reference as a baseline for Figure 8.1 in the manuscript, and modified the milestone descriptions based on personal experience and application to military aircraft programs.

2. Joint Publication 1-02, Department of Defense Dictionary of Military and Associated Terms, 8 November 2010 (As Amended Through 15 October 2015) http://www.dtic.mil/doctrine/new pubs/jp1 02.pdf (Retrieved on November 5th, 2015)

3. Gunn, Thomas, *Gunnsights*, pp. 93-102, Annapolis, MD: Naval Institute Press, 2007

4. Work Breakdown Structure (WBS) example (Figure 7.1):

- Level 1 in the WBS is the aircraft weapon system.

- Level 2 teams include Air Vehicle, and Support and Training,

- Level 3 teams under Air Vehicle include the Fuselage, Wing, Empennage and Subsystems.

- Level 4 teams under the Wing IPT include the wing torque box, control surfaces, fixed leading edge, trailing edge and wing tip structure.

- Level 5 examples under the Control Surface team include Flaps and Ailerons.

CHAPTER NINE
MANAGING COMPLEXITY

Fools ignore complexity. Pragmatists suffer it. Some can avoid it. Geniuses remove it.

Alan Perlis
American Computer Scientist

E arlier, I described the "cradle to grave" process for aircraft programs—how they are designed, built, tested, fielded and supported over many decades of operational use. For each of the two business units where I led engineering, there was a portfolio of programs, all in different phases of the Product Life Cycle. Each phase or stage of development involved different challenges, considerations, decisions, and people with skillsets and strengths associated with design maturity. Managing the complexities associated with such diverse programs in an ever-changing global environment was both a challenge and a core competency for leaders. In addition, the programs included customers from all branches of the United States and allied military forces. Further, these programs were being performed at various geographical sites each with its own culture.

There were many dimensions to complexity, but it was my job to manage it, simplify it and even use it as a competitive advantage.

Table 9.1 defines the complexities that required knowledge and focus when I was chief engineer for Support and Services, and the Military Aircraft Businesses.

Table 9.1 Managing and Simplifying Complexity

SOURCES OF COMPLEXITY	KEY DIFFERENCES EXAMPLES	SIMPLIFY AND MANAGE (STEPS TAKEN)
Different Product Lines • Fixed Wing Aircraft e.g., Fighters, Trainers, Bombers • Rotorcraft e.g., Tilt Rotor, Tandem Rotor • Commercial Derivative e.g., Tankers, ISR missions • Weapons • Unmanned Vehicles	• Program Life Cycle Phases e.g., development, production, support • Fixed Wing vs. Rotorcraft • Boeing vs. Non-Boeing as OEM	• Organized to manage this level of complexity • Deployed technical leadership and engineering talent across the business portfolio. • Implemented an Engineering System to drive First Time Quality • Utilized Reverse Engineering processes for Non-Boeing platforms • Initiated Improved Supplier Technical Oversight Process
Spectrum of Mission Capabilities • Strike/ Attack • Mobility • Intelligence, Surveillance and Reconnaissance (ISR) • Aerial Refueling • Modifications and Upgrades • Support and Training	• Customer's Most Important Requirements • Design Criteria • Mission Avionics and Equipment • Level of Technical Integration • Airworthiness Certification Processes	• Implemented disciplined Systems Engineering processes • Identified Domain Knowledge requirements and hired aggressively • Shared Best Practices and Lessons Learned

Employee Demographics		
• Retirements increased each year • Stagnant growth in STEM (Science, Technology, Engineering, Mathematics) college graduates • Fewer Development Programs for engineers to work	• Site Capabilities and Capacities • Common staffing system to track emerging and future needs	• Cored up "selected" resource capabilities into Engineering Centers of Expertise (chapter seven) • Implemented Technical Mentoring and Knowledge Transfer approaches • Moved work packages between sites consistent with capability and capacity.
Multiple Heritage Companies, Sites and Cultures		
• Became One Company Through Acquisitions and Mergers • Different Organizational Models • Different IT Systems and Business Processes	• Design and Build not always at the same location • Level of Supportability and Safety considerations in designs integration in Product Teams • Degree of Design for Manufacturing and Assembly,	• Provided and communicated a vision for the engineering organization • Engaged "locally" with people and teams • Relocated key talent across different sites • Implemented common processes and tools • Performed capability and capacity assessment across the sites
Global Environment		
• Broadening global threats • Geopolitical considerations • Sequestration; Reduced DoD spending levels • Projected numbers of students graduating with Engineering Degrees and entering the workforce in 2020.	• Fewer, but critical, new franchise programs • Shifting Risks From Cost Type Contracts to Fixed Priced Contracts on Development Programs • Continued Industrial base restructuring	• Pursued International Opportunities to offset less DoD spending • Focused R&D spending to address customer's emerging needs • Positioned the engineering team to align with business vision and strategy • Promoted STEM Education Initiatives

The "Sources of Complexity" refers to the breadth of platforms. There were significant differences in types of aircraft, which required specific domain knowledge and technical expertise. In the second column, "Key Differences" highlight that the programs could be and were in different phases of the Product Life Cycle. Another dimension: was Boeing the Original Equipment Manufacturer (OEM) or not, and how were the different aircraft certified for airworthiness? The third column, "Steps Taken" (to manage and simplify the complexity), lists some of the actions taken under my leadership. Some of these actions are explained in more detail in this chapter.

Organized to Manage Complexity

When I began my assignment as chief engineer in Boeing Military Aircraft (BMA), there were four profit and loss divisions. Each division had a portfolio of programs, each program with a chief engineer or director of engineering, who was aligned with the division Vice President/General Manager (VP/GM) for day-to-day direction. While the alignment was effective for tactical tasks and strategy development and deployment, there wasn't as much integration or sharing of best practices, applying lessons learned and resource sharing between the division chief engineers as there could have been.

Since each division had development and production programs, we changed the organizational structure at the business unit level and assigned one chief engineer for development programs and another for production programs. With this type of structure, we were able to share best practices and lessons learned across the entire business unit. This alignment also enabled our team to mobilize and deploy expertise from both within the BMA engineering team and across the enterprise, to ensure the right engineering   talent was brought to bear on specific needs. Having leaders focused on either development programs or production programs helped simplify and manage complexity across the organization.

I felt this was a very effective and productive model. Knowing capabilities and strengths of people and teams on various programs located at different sites, the BMA engineering team was able to deploy resources between programs to bring expertise required either to help solve problems or conduct peer reviews. Further, it was a great way to give people cross training. For example, subsystems engineers assigned to rotorcraft programs were deployed to commercial derivative programs. My team used this practice very effectively and helped the collective BMA organization execute programs.

Deployed Common Processes and Tools

3D Modeling and Analysis Tools. Once McDonnell Douglas and Boeing became one company, it was a key focus to drive toward common tools for design and analysis. Obviously, trying to work with multiple systems—paying license fees, conducting maintenance, and training engineers—was cost prohibitive. Having a common set of tools enabled engineers to move from program to program and even site to site with a high level of productivity, and it also enabled the transfer of work between sites, to maximize capability and capacity.

Through the Enterprise Engineering team and BDS engineering functional organization, many tools (and associated cost) were eliminated. The company did, however, maintain two 3D modeling tools for three reasons. First, there were so many products still in production which had been modeled in different systems, and the cost of conversion would be prohibitive. Second, engineers were trained and proficient in the systems they used. Third, there were adequate software tools to facilitate interoperability between the two 3D systems.

Lean+ Process Deployment. During my tenure as the chief engineer in Global Services and Support, in the latter half of the 2000s, the Defense, Space and Security chief engineer would call all engineering managers to an offsite in Newport Beach, CA.

A review of facts and data on cost and schedule overruns on Department of Defense programs convinced us not just that we needed to improve, but had to develop a process to change how our engineers approached their work. The key tenets: (1) prioritize work, (2) focus and finish tasks, and (3) eliminate multitasking (having too much work-in-process ongoing) with a goal of first time quality engineering.[1]

It fell to my fellow engineering managers and myself to lead the implementation of this new approach across our respective business units. I traveled to each site where my unit performed work and held all-day meetings with engineers and product support professionals. We replicated the agenda that created the compelling need for change. With video of Boeing products in action to start the day off on the right foot, we would use a combination of presentation material, hands-on training, and open dialogue to facilitate a very productive day.

Within three months, we began to see the fruits of our labors, Teams measured improvement in quality and throughput and created "Work-in-Process" status boards. I made it a point to spend a couple of hours a week visiting work areas, talking with teams about their accomplishments and challenges with implementing this new approach.

These Lean+ techniques were especially effective when applied to tasks such as drawings releases, software development, manufacturing work instructions, and technical orders. This proven process had a track record of positive results in both productivity and quality. At my business units, we measured quality, productivity (or "throughput improvement"), and bottom-line cost savings. In the aggregate, we would see between 3X and 5X improvement in quality and throughput.

Leveraging Lean+ processes created the capacity for our teams to focus on innovation and growth. In the beginning, we held weekly meetings for program chief engineers and their "Lean" team leaders to review implementation, including cost reduction results vs. target. These meetings connected people who normally would never talk to each other. By bringing the program chiefs together, we facilitated exchanges of information and assistance. If, for example, the C-17 program was briefing on success in applying Lean+ practices to wire harness design, routing and installation, and the F-15 program was having a similar problem, the two programs could get together after the meeting to discuss—a great example of one program helping another. Lean+ is an excellent process for the engineering toolbox.

Root Cause and Corrective Action (RCCA)

During program "define, build, test, and support" phases, issues would arise, along with opportunities for continuous improvement. They would, of course, be addressed by program management, but from one program to another, approaches might differ widely. When we formed the new Boeing Military Aircraft engineering organization, the chief engineer for Production Programs brought along a very robust process for getting to the root cause of a problem. This involved getting answers to four questions. "What do you **K**now, what do you **N**eed to know, what are

9.2 Premature Failure of a Structural Component during Vibration Testing

Data Element	Know K	Need N	Opinion O	Think T	Action Items
Failure occurred at an area of high stress concentration[2] during vibration testing prior to completing the planned number of cycles.	X				N/A
The Structural modeling and analysis was peer reviewed. Technical assumptions were consistent with company and industry best practices.	←——X				Conduct an independent review of the technical approach, analysis and verify that key assumptions are consistent with company and industry best practices.
A flaw in the material alloy caused the failure. The machining process during part fabrication caused an unexpected stress raiser.	←———————X				Perform a laboratory analysis of the failure surface for failure location and failure mode. Check material hardness and other material properties to ensure the material alloy was processed correctly.
The vibration loads were correctly applied. The test fixture was built as-designed and the structural component was properly installed in the fixture to represent the aircraft installation.	←——————————— X				Ensure that design vibration loads requirements were flowed down with clarity to the responsible team and properly understood. Verify that the test set-up was consistent with the test plan. Review the data from the test to ensure the vibration levels were applied correctly during the test.

Opinions, and what do you Think you know?" A KNOT chart is an effective template for root cause analysis, and became a required component of RCCA. Table 9.2 depicts an example KNOT chart and it is used as a both a way to organize facts and data, and as an action plan

The biggest challenge was getting teams to take the time and work through the process. It was human nature for people to go from what they thought was the cause of the problem to devising what they thought was the solution. KNOT represented a simple, straightforward, and systematic process leading to a comprehensive understanding of the root cause of a problem. Once people saw the process in action, they became believers.

Based on my experience, the conclusions from the collective actions will likely identify the root cause or causes of the premature failure. Corrective actions can range from correcting the test fixture and repeating the test to a part redesign, and re-test...and believe me, you don't want to go to the latter scenario.

Earned Value Management (EVM)

One of my previous leaders would talk about "The Value of a Day." For example, on average, a development program takes approximately 60 months, from Contract Award to completion of Development Test and Evaluation. If you think about, oh, 1000 people working for 60 months, you might want to know how much work needs to be accomplished each day and each week to keep the program on schedule and on budget. Earned Value Management is a process with an information system designed to ensure that IPTs are completing the right amount of work within the planned resource level.

Imagine, for example, what might happen if a program team comes to the 60-month milestone and discovers, oops! They still have eight months of work yet to complete. Which means, of course, delivery will be at least eight months late and significantly over-budget. With EVM in place, red flags would have been raised long before the point of crisis. Program management, chief engineer, and the IPT leaders would have been engaged, to explore and correct.

Built Strong Customer Relationships

Boeing had field offices at major military bases, and I always made it a point to visit customers at their facilities: Wright Patterson Air Force Base, Ohio; the three Air Force Air Logistics Centers (ALC)—Oklahoma City, (Tinker AFB), Ogden, Utah and Warner Robins Georgia—and the Army's Redstone Arsenal in Huntsville, Alabama. I would meet with customer chief engineers, program managers, and those leaders and teams responsible for airworthiness certification. I felt it important enough to hear first-hand what our customers thought about Boeing engineering teams across the board, and to take action on their feedback, as appropriate. Of course, I wasn't flying blind, so to speak: thanks to the outstanding support of Boeing's Field Offices, I already knew about most of the customer issues and was ready to respond. In addition, I made sure that issues and actions were not only acted on, but that the information was passed along to other Boeing leaders who might later be visiting the customer.■

Chapter Nine Technical Takeaways

• Organizational structure, a robust product development process, use of common processes and tools, and the right level of domain knowledge and expertise—

steps along the path to managing complexity.

• Make it a priority to KNOW your customers and communicate with them often. Obtain feedback and use it as a means to improve performance and or improve customer satisfaction. Document and follow-up on all action items.

Chapter Nine Leadership Takeaways

• Managing and simplifying complexity is a core competency for leaders in the 21st century.

• As a chief engineer responsible for a portfolio of complex programs, you have to be able to manage your time (1) to focus on high priority programs, (2) to position the business for the future, and (3) to understand the many aspects of people. Your success is dependent on having the right team in place, at the right time.

• The ability to lead change such that people embrace it is also a key competency for leaders today. Be a role model for change, teach it, expect it, and reward it.

---

1. Boeing Frontiers Magazine, October 2009 [Lean+] Simple As, pp. 36-38 http://www.boeing.com/news/frontiers/archive/2009/october/i_ids01.pdf (Retrieved November 12th, 2015)

2. Areas of high stress concentration or stress raisers are holes, grooves and sharp corners. Stress concentrations can increase stresses locally by a factor of approximately two (2) for a groove and a factor of three (3) or more for holes.

CHAPTER TEN
SELECTED CASE STUDIES

The Only Easy Day Was Yesterday
US Navy SEALS

T he first five Case Studies described in this Chapter span 14 years and two Business Units, Boeing Military Aircraft (BMA) and Global Services and Support (GS&S). They highlight a cross section of the multiple roles and responsibilities I had, which included eight years as a Business Unit Chief Engineer, and represent the challenges and setbacks that provide the greatest opportunity for learning. Each describes a specific program and situation, the actions and decisions I made or was involved in, and the outcome, key best practices, and lessons learned. The sixth Case Study highlights the complexities and challenges with a DoD development program outside of aerospace.

While these two business units had similarities and differences, to some extent, they were inextricably linked. Global Services and Support supported many programs developed and built in Boeing Military Aircraft. BMA had much fewer contracts than GS&S but with much larger contract values. Both organizations would have presence at multiple sites where work was performed. The customers and global environment would drive different business strategies, cost structures and investment priorities.

Military Aircraft Business

Background. I was named vice president of engineering for Boeing Military Aircraft (BMA) in October of 2010 and held the position until my retirement in November in 2013.

I was responsible for engineering across a portfolio of programs, including the Italy Tanker Program, the Airborne Early Warning and Control (AEWC) Program for Australia (also known as *Wedgetail*), development of F-15SA (newest Saudi Aircraft model), the U.S. Air Force KC-46 aerial refueling tanker, the P-8A *Poseidon* for the U.S. Navy and India, and multiple *Chinook* international programs (United Kingdom, Netherlands, Canada and Italy). Plus, "Scan Eagle," an unmanned platform which was developed and built by Insitu, a wholly owned subsidiary of Boeing.

This was an incredible portfolio, with contracts valued in the billions. Some of these programs were commercial derivatives, that is, commercial aircraft modified for military use. The Italy Tanker aircraft were Boeing 767s equipped for aerial refueling; the P-8A *Poseidon* and AEWC aircraft were Boeing 737s modified for Intelligence, Surveillance and Reconnaissance (ISR) missions.

I came aboard at a critical time; the military aircraft business was faced with great change. Restructuring and mergers were reducing the number of competitors, the DoD was shifting more risk to contractors by moving away from "cost-plus" to "fixed-price" contracts, and fewer "new" programs were on the U.S. horizon— putting a focus on international business.

The president of Boeing Military Aircraft created a new organization—the Operating Executive Team—which integrated the

Engineering, Supplier Management and Procurement, and Operations and Quality functions. A very experienced, senior and visionary executive was assigned to lead this team as the Operating Executive (OE) or Chief Operating Officer of BMA. The Charter of the Operating Executive team was to:

- Focus on improving program execution by working together with the programs. This included:

  ○ Identify the root causes of schedule delays and cost overruns on development programs, and determine and implement corrective action

  ○ Help bring challenged programs to closure sharing best practices and applying lessons learned to programs beginning or in development.

  ○ While program organizations had the responsibility, accountability and authority (RAA) to execute their respective programs, they would not be the only ones being held responsible for execution.

- To differentiate Boeing Military Aircraft from competitors and position the business for a future with fewer franchise programs, we would develop and implement strategic initiatives, along with game-changing ways of doing business, to more rapidly deliver capabilities to customers. This could be a key competitive advantage.

Driving Execution through an Integrated Organization on Key Development Programs

Concurrent with the formation of the Operating Executive Team, I had work to do, establishing the engineering leadership team.

When I came into the organization, we had chief engineers providing support to each of the four operating divisions—Global Strike; Weapons and Unmanned Systems; Vertical Lift; and Mobility, Surveillance, and Engagement. They reported to the VP/GMs of the respective divisions, and to the functional organizations at each site, providing support to programs and managing process and tool developments. To support the charter of the Operating Executive's team, we needed an organization with a collective focus across all of BMA on technical excellence and situational awareness of our programs. Given that "structure follows strategy," I realigned my direct engineering staff to focus on development programs, production programs, and functional support, rather than aligning with the four divisions within BMA.

Our Operating Executive organization got rolling in the fourth quarter of 2010. We would spend the first 90 days getting focused on the priorities, building relationships across the BMA organization, and begin working with the programs. Two of them were, well, challenged, and required special focus: the Tanker Program for Italy and the Airborne Early Warning and Control (AEWC) Program ("*Wedgetail*") for Australia.[1]

Both programs had technical issues, largely, some aspects of the mission systems didn't meet contract requirements. Those issues in turn caused late deliveries, unhappy customers, and cost overruns. The Operating Executive team brought a systems level approach to solving these and programmatic problems. We built relationships, not just with the program, but with customers and the industry teams, which included the supply chain. We facilitated comprehensive root-cause analyses, which help teams determine "what is known" and "what needs to be known" to effectively resolve the problems. Action items naturally flow out of this process, and the OE team took on as many action items as were appropriate to help offload the teams. We poked at design assumptions on both programs and studied to what degree those assumptions were anchored with analysis and/or test data. Where necessary, we would add resources with the right expertise to help drive daily execution. Resources with specific expertise were assigned to augment the program team until the technical problems were resolved. The OE team stayed actively engaged together with both programs until all aircraft were delivered.

The first Italy Tanker was delivered at the end of 2010. The following three aircraft were delivered in 2011.

Figure 10.1 Italy Tanker aircraft in flight (Copyright, Boeing)

On the *Wedgetail*—Boeing's site in Seattle completed the first two aircraft modifications. The remaining four were modified in Australia at Royal Australian Air Force Base (RAAF) Amberley. BAE Systems in Australia supplied the electronic support measures and electronic warfare self-protection systems; requiring significant development work with an Israeli supplier on the former to integrate into the aircraft.

Figure 10.2 AEWC Aircraft Takeoff (Copyright, Boeing)

We would facilitate needed coordination, and provide leadership and technical expertise to the supply chain. Together with the program and onsite leadership, the Operating Executive Team collectively helped develop and implement action plans to successfully complete development and deliver all six aircraft to the Royal Australian Air Force in 2012.

The Operating Executive would always tell us "Know the difference between effort and results."[2] While, working together with the programs, we got results . . . but "effort" was certainly a component. We traveled the world to get those "results," many times.

A cockpit upgrade for the already operational United Kingdom's Royal Air Force fleet of 38 Mk2/2a and eight Mk3R *Chinook* helicopters was in work. Further, a contract was in place to procure 14 new Mk 6 *Chinooks*. The Operating Executive team would help work through technical issues with the in-country supplier (See Case Study on U.K. *Chinook* Programs).

Figure 10.3 Boeing 767-2C conducted its first flight on December 28th, 2014. (Copyright, Boeing)

The KC-46 tanker was working through the left hand side of the Systems Engineering V. The KC-46 tanker is a provisioned Boeing 767 aircraft modified for military use, designated as the 767-2C. This configuration included design and installations of four auxiliary fuel tanks inside the fuselage, and structural and systems attachments or interfaces for the refueling boom, the wing aerial refueling pods (WARP) that attached to the WARP pylon, a centerline drogue system (CDS) and other mission equipment. The weapon system modification that bring the 767-2C to a KC-46 include aerial refueling mission equipment and the designs to integrate them into the aircraft through physical, functional and software interfaces.

The first flight of the KC-46 was on September 25, 2015. As of this writing, the program EMD aircraft are in flight-testing.

Positioning Engineering to Support the Business for the Future—was a key responsibility of my job as a Business Unit Chief Engineer. Given the changing global environment, it was imperative that my organization be aligned with any and all future decisions or changes in direction, which included:

1. To ensure solid execution on our existing contracts, especially those of highest visibility to customers and the corporation. Executing current contracts to plan is a key strategy for business growth.

2. The larger and more complex development programs were fewer and far between. As a result, keeping our resources productively employed was critical; we looked for opportunities, both within and outside BMA, to shift work between sites, even business units like Commercial Airplanes, depending on demand and availability. Further, it was to ensure keeping a critical mass of resources aligned with those areas of engineering considered "must design," in-house.

3. It was important to stay in front of future customer wants, needs, and requirements.

a. As part of the source selection process, some programs on the horizon would require prototype aircraft demonstration of key capabilities. My role with the programs was to select and deploy the right engineers, especially those with the experience and attributes for the up-front or technology phase of the program. In addition, I would engage at strategic points in development to review program progress, offer appropriate guidance or feedback and address help needed, whether technical input and or to provide specific skilled people.

b. The Boeing Company has an Enterprise Technology Process for our Research and Development (R&D) investments. My responsibilities included integrating feedback from all stakeholders on R&D projects. This ensured they were prioritized and aligned with both future customer needs and business plans. The final step was a review with both business unit and engineering functional leadership prior to submission to the chief technology officer.

4. To "stay in touch" with our customers, we made a visit to the chief scientists of the respective military services. They shared feedback on their activities and concerns, which allowed us to ensure that our R&D investments were aligned with our customers' thinking and direction.

Case Study: F-15SA (Saudi) Program Fly-by-Wire Flight Control System

Developing a Major Technology Upgrade to a Legacy Aircraft

In December 2011, the Kingdom of Saudi Arabia purchased 84 new F-15 aircraft, designated as F-15SA. This was contracted as a Foreign Military Sales (FMS) program, with the U. S. Air Force as the agent. The development and production of this new model fell to Boeing Military Aircraft and the F-15 program engineering team. Global Services and Support had responsibility for follow-on support.

The F-15SA is a variant of the F-15E, with several upgrades including a fly-by-wire flight control system and new mission avionics. The F-15SA has two F-110-GE-129 engines, each of which produces almost 30,000 pounds each of thrust, 60,000 lbs. total.

As a business unit chief engineer, I would engage with program chief engineers and IPTs that were early in development and at a frequency—sometimes once a week—that depended on program phase, engineering complexity, and criticality of achieving key milestones. Engaging with people was one of the best parts of my job. It was a great opportunity to really understand the scope of development, determine if we were taking the right technical approach, and if we were on track to complete on time and within budget.

During one of several deep-dives on the Fly-By-Wire flight control system, it was clear the team was challenged and was struggling to meet the schedule. In fact, they were on a path to missing some key programmatic milestones.

As described in Chapters Four and Eight, and depicted in Figure 4.3, "flight control" is an integrated system which translates pilot stick and rudder pedal inputs into pitch, roll and yaw commands, characterized by load factors (pulling g's or rolling/pitching the aircraft at a specified roll/ pitch/ yaw rate or acceleration). Flight control design depends on robust system architecture, coupled with aerodynamics from wind tunnel or flight test for the operational envelope (speed, altitude, angle of attack, and other aerodynamic parameters). To ensure that structural fatigue and damage tolerance requirements are met, flight control and structural engineers need to understand what and to what degree flight control surfaces—ailerons, flaps, horizontal stabilators, and rudders—will be used over the lifetime of the aircraft.

It was clear the F-15SA team was in need of executive engineering leadership with expertise in aerodynamics and flight con-

Figure 10.4 F-15 Saudi Aircraft (SA). First Flight, February 20th, 2013. (Copyright, Boeing)

trol law development. The first step: to build a plan to test and field incremental capability. The first build would be of limited capability but would enable the program to achieve first flight. More important, though, it was a confidence builder, to demonstrate that our system architecture was technically sound. Next, we brought in some outside resources, specifically subject matter experts—including retired engineers—in flight control system development. The breadth of issues identified by the independent experts was significant. We took a methodical approach to address each and every finding, while we continued on our path to develop, test and field incremental capability.

One area in need of improvement was an automated analysis tool—engineers who design and test products depend on engineering tools, much like mechanics and electricians depend on other tools to build product. When tools are not effective or efficient, productivity suffers, and this particular tool was labor-intensive and took too long to do the job.

I contacted our functional team responsible for tool development; the team quickly shifted priorities and began work on a corrective plan and gave me a progress report on a weekly basis. I held them accountable to address in a timely manner, and they did. This kind of engagement by a chief engineer helps the team surface issues, but also ensures that appropriate resources will be brought in, as necessary.

With the team leadership and resources in place, we would complete first flight only a few months late to schedule . . . a greatly improved, although not all that commendable, result. Concurrent with collecting data from the initial flights, the team proceeded on developing the next block of capability, which opened up the envelope to begin flight-testing in October 2013. Ironically, I would see the three test aircraft depart St. Louis while in the airport a couple of weeks after my retirement.

Also, I came to realize that some changes in organizational structure might be in order for flight control systems development. This particular skillset of flight engineering had the smallest population. Further, there were multiple flight control law methodologies, processes, and tools that could be utilized to complete development. The decision on which method to use would be determined by the engineers in place. I felt with this level of technical complexity and the importance of first time quality execution, more structure and process discipline would be in order. As a result of this experience on the F-15SA program, I pushed to create a Center of Expertise for Fixed Wing Flight Sciences. The model for this organization would be based around Boeing Military Aircraft Flight Sciences work performed for Vertical Lift or Rotorcraft programs, which had worked very well. Key objectives and advantages of a Center of Expertise were (1) a disciplined approach to establish the appropriate flight control law architecture for a given design application, (2) knowledge transfer between the more senior engineers and junior engineers, (3) sharing best practices and lessons learned for development, and (4) consistent technical training. This model was put in place shortly after I retired in late 2013.

Key Case Study Takeaways

● As a chief engineer, engage frequently with teams during development phase of programs. Listen and look for opportunities to assist teams in need of help and provide it in a timely manner.

● Ensure scope, schedule and resources are aligned so the team has an executable plan and can be successful.

● Continually assess technical risks, and ensure steps are being taken to mitigate them.

● Take necessary actions to ensure root causes are understood and corrected so they don't recur on future programs. Creating an Engineering Center of Expertise in this case was the right step to take.

Case Study: P-8A Fatigue Test

Ensuring Technical Integrity on a Commercial Derivative Aircraft

Background for the Case Study: One of the roles of an executive leader is to step up and take on tough situations at the enterprise level—that is, dealing across the Corporation, not just within the assigned business unit or division. In this particular case, the P-8A military aircraft program and the 737 commercial airplane program would be competing for the same facility space at the same time. (The P-8A is a derivative of a 737-800 aircraft.) The P-8A program needed the space to conduct and complete a structural fatigue test, which had a recent change in work scope (and schedule). The commercial program wanted the space to increase production rate.

If this is the case, you might ask, "Why do we need to conduct a fatigue test with all the flight hours already on commercial 737-800 aircraft?" The missions for the US Navy are anti-submarine, anti-surface and intelligence, surveillance and reconnaissance (ISR). After takeoff the P-8A will fly at speeds and altitudes—an operational envelope—much different than a 737 commercial airliner. The P-8A also carries weapons under the wings and the interior of the airplane is significantly different. Therefore, because of the structural changes and a loads environment and fatigue spectrum that is so different from the 737-800, a fatigue test was required.

Case Study: The military program customer wanted to extend the testing from the initial contract requirement of 2 lifetimes of testing to 3 lifetimes of testing.

We know that the military services usually fly their aircraft longer than originally planned, supporting the extensions with numerous upgrade and modernization programs. Here, we had the customer thinking ahead, call it, "be prepared." This situation, however, was complicated because the request to test the 3rd lifetime came after the contract called for testing for only 2 lifetimes. The additional lifetime testing would take approximately 18 months to

Figure 10.5 P-8A Aircraft with Weapons Bay Doors Open (Copyright, Boeing)

complete.

The P-8A program wanted to keep the test aircraft in the same physical space for all three lifetimes, forcing the commercial airplane program to come up with another plan for increasing their production rate. The commercial program wanted the military program to "surrender" the space after 2 lifetimes, then relocate somewhere else for the third lifetime testing. What next? The company's chief technology officer, after consultation with senior leadership of the Military and Commercial units, would make the decision.

To this point, I had not been involved in the issue, but I stepped up and offered assistance. The program was within BMA's portfolio, and the problem was within my technical area of expertise.

Going in, my gut feeling was that we simply could not move the airplane until all testing was complete. However, engineers make technical decisions based on facts and data, and not gut feelings.

Consequently, I became the team leader with a really great executive from the Commercial Airplanes programs as co-leader. He had a strong structures background but also had the responsibility to increase the production rate.

As you would expect, we pulled in a group of experts from both the military and commercial side of the business. This included structures senior chief engineers, facilities people, and the test team which could address the risks of moving a full-scale test article and resetting it up. In other words, (1) could we move the aircraft between the second and third lifetime and (2) did we have the capability to set the aircraft back up with all load cells and be able to replicate the applied loads, deflections, stresses and strains? We studied plans, schedules, options, and risks.

A key factor: to study the facility where the test article was sitting. The floor had been strengthened with thousands of pounds of re-bar or reinforced concrete to stabilize the airplane. The test setup was very complex and included an in-place control center, which provided real-time monitoring of strains and deflections; it would have been difficult to replicate. Also, towing the airplane out of the facility might have caused something to shift ever so slightly . . . but that could be critical. There was no other site in the local area that had the reinforced floor and all the additional equipment required to conduct such a complex test.

After a team evaluation, it was determined that the risk of moving the aircraft between the 2nd and 3rd lifetime of testing and replicating the test setup was too high for the P-8A program. I gave big credit to the Commercial Airplanes team lead for finding an option, another space, that met the 737 program's production rate requirements. It would be the right decision and win-win for both sides of the business. This exemplified a One-Boeing approach.

Key Case Study Takeaways

- The right team in the right place at the right time, working together and wearing their "enterprise hat" (as noted, "across the Corporation") will make the right decision.

- Customers are flying aircraft longer than ever intended. Plan ahead . . .

Support and Services Business

Background: Traditionally, once aircraft were fielded, customers were responsible for sustainment, logistics support and mission readiness of their fleet. However, through the Product Support organization, we provided: (1) technical data for the maintainers to conduct troubleshooting or maintenance, (2) systems support analysis to determine the quantity of spares customers would need, (3) design and production of retrofit kits for change incorporation into aircraft, and (4) put in place the supply chain to repair parts. Customers generally tasked the program engineering team for fleet support, especially if the aircraft weren't performing as expected. Usually, flight and maintenance training devices were provided through the McDonnell Douglas Training organization because in most cases we were the Original Equipment Manufacturer (OEM) of the aircraft. However, customers would sometimes put training devices out for bid, giving other companies the opportunity to develop, for example, the simulator for an aircraft we designed—and vice-versa. The whole business of "support" could be fragmented and in some cases, our response might not be up to the standards we expected of ourselves. In the latter half of the 1990s, a new Military Aerospace Support

Division was formed, to bring all logistics support and training under one organization. The rationale:

1. Approximately 70 percent of the lifetime cost of a weapon system—maintenance, modifications, up-grades, spares, repairs, kits, training devices, ground support equipment and technical updates—was incurred after the product was delivered.

We were looking for ways to grow our own business by winning more of that 70 percent chunk.

2. By better tailoring business models and solutions to customer needs, Aerospace Support could improve mission readiness.

3. On the customer side, the Government was work-ing through a Base Realignment and Closures (BRAC) process to realign assets to changing needs.[3] The KC-135 maintenance depot at Kelly Air Force Base in San Antonio, Texas was being closed—but the maintenance requirement remained, to be shifted to commercial sup-pliers. Aerospace Support took on the Programmed Depot Maintenance (PDM) for the KC-135.

4. Customers wanted to modernize and upgrade capa-bilities to keep some of their platforms flying longer. Further, customers were competing these modifications and upgrades when programs were out of production. As a result, a company like Boeing could compete to maintain, modify and upgrade a platform of which it was not the OEM. Examples that highlight success stories and associated challenges with performing this kind of work are included in the three Global Services and Support case studies later in this Chapter.

5. Because "services" can generally be provided with

fewer fixed assets—plant, property and equipment (PP&E)—than manufacturing programs, they offer a higher level of economic profit. Examples that highlight success stories and associated challenges with competing for support contracts are included in the following case studies.

6. This increased emphasis on services and support put a focus on "Design for Supportability and Maintainability" up front in the process. The quality of engineering would directly affect the maintainer's ability to support the aircraft.

Within a few years,, Aerospace Support would have a diverse portfolio of some 10,000 contracts—commercial derivative aircraft, fixed wing aircraft, and rotorcraft programs—on both Boeing platforms (OEM) and non-Boeing platforms. Work was being performed at multiple sites, domestic and international.

This was an incredibly complex, yet exciting business, given the diversity in programs and cultural differences at the various sites. These contracts ran from thousands to billions of dollars. Business ranged from purchasing parts (transactional) to payment for a service known as a Performance Based Logistics (PBL) contract (where customers are paying for a level of mission readiness). While PBL contracts are structured around the needs of specific customers, they generally require programs to manage supply counters at operational bases, manage repairs at supplier's facilities, provide engineering support and perform depot level maintenance and modifications on aircraft. Our Engineering, Logistics and Operations work would be performed predominately at five domestic sites and three international sites (the United Kingdom, Australia, and Saudi Arabia). In some cases, engineering for capability upgrades would be performed at one site, with modification work at another, which added a level of complexity.

I spent 11 years of my career in Aerospace Support. For my last four plus years, I was Vice President and Chief Engineer. The organization's name changed twice, most recently to Global Services & Support, to better reflect the business vision. There were five presidents of Aerospace Support over that time-frame. Each brought their own leadership style, priorities, and areas of focus; the staff would adapt and adjust. I managed approximately 3500 engineers, with a dual responsibility for almost 4000 Product Support professionals across BDS. However, at the end of the day, my job was about the customer, ensuring technical excellence and results. Further, it was about finding ways to best integrate the engineers and logisticians, to provide supportable and maintainable aircraft. In other words, designing for supportability was just as important to the warfighter as designing for manufacturing was to the operations team during production.

The following case studies highlight the work, my involvement as a leader, and some of the challenges I faced. In addition, I share "best practices and lessons learned" as the services and support business continued a trajectory of both top line and bottom line growth, through the first decade of the 21st century.

Case Study: T-38C Talon

Upgrading the Entire Fleet of Jets that Train All United States Air Force Pilots Who Will Fly Fighters and Bombers

I joined Aerospace Support in September of 1999, as the T-38C Avionics Upgrade Program manager. The organization didn't have a chief engineer, but it had four engineering-centric IPTs reporting directly to me. So, *de facto*, I was chief engineer, integrating the efforts of these four teams. With responsibility for the entire program, to include trainer or simulator development, technical data for maintenance, and supply chain sustainment IPTs, I wasn't sure it was in the best interest of the program long term to dual-hat as the chief engineer. With my background however, and with the technical problems on the program, my decision was to do just that, at least initially. This way, I could better assess the program's technical challenges, relationships with the customer, team leadership and overall talent on the team.

The T-38C was, and still is, the aircraft that trains all U.S. Air Force pilots who will fly fighters and bombers. Northrop Grumman was the OEM for the aircraft, which had been out of production for many years. The T-38C upgrade removed the analog cockpit and replaced it with a digital glass cockpit, similar to a cockpit that fixed wing pilots fly in aircraft such as the F-15 and the F-22. The upgrade would enhance continuity in pilot training. There would be an opportunity to upgrade as many as 519 aircraft, with most of these aircraft within the Air Education and Training Command (AETC), headquartered in San Antonio, Texas.

This program was different than any program awarded to Boeing at the time, but aligned very well with the charter of and why the Aerospace Support organization was formed.

> 1. Boeing was not the OEM. While we had expertise in developing digital glass cockpit avionics and software, we would have to rely on the customer for technical information in order to incorporate and integrate avionics, structural, mechanical, and electrical changes.

2. The Air Force was under a mandate to train a given number of pilots per year . . . and was struggling, because of reliability issues and too-frequent maintenance. When the Air Force contracted with Boeing to develop, modify, and support a cockpit avionics upgrade, they imposed a Mean Time Between Failure (MTBF) requirement of 60 hours. That meant, on average, that any aircraft could only have one maintenance action for every 60 hours of flight time. This Key Performance Parameter (KPP) would be tracked at many levels of government and within Boeing leadership. I knew that meeting this KPP was going to be a big challenge.

3. Boeing would be responsible to operate "Contractor Supply Counters" at each of the six AETC bases, making parts readily available for replacement. Response times for supplying parts were defined in the contractual requirements.

4. When we had finished our in-house development and testing, aircraft were taken from the fleet and flown to our modification center in Mesa, Arizona. The target: as many as seven completed modifications, per month. There wasn't much room for missing the target; the Training Command would tell us "The replacement for the T-38 was the T-38," meaning they expected that upgraded aircraft would be sent back, as quickly as possible.

5. Concurrently, Boeing was contracted to develop and deliver aircrew training devices and instructional courseware to augment pilot training.

6. Technical data was also a contract deliverable; the maintainers would be required to effectively troubleshoot problems and keep the aircraft flying and pilots

moving through the pipeline.

My job was to integrate the efforts of engineering, finance, contracts and pricing, supplier management and procurement, the modification team, logistics or product support, trainers and training. (That is not a laundry list, that was my job!) We engaged an Israeli company as prime subcontractor/partner; some years earlier, they had performed a cockpit upgrade on F-5 aircraft (the T-38 is a derivative of the F-5) for a foreign customer. We could help reduce cost by adopting a large percentage of their software from that previous customer. We also planned to leverage "Commercial Off-The-Shelf" (COTS) and Non-Development Items (NDI)—items that had been developed for other platforms.

I was the second program manager (PM) to lead the T-38 Avionics Upgrade Program. Shortly after taking on the assignment, we received our Contractor Performance Assessment Report (CPAR). It was not flattering. There were un-met challenges. We were scored "marginal" in many technical categories—that is, somewhere between "satisfactory" and "unsatisfactory."

Thus it was that, approximately six weeks later, I found myself under fire at the Pentagon. A Saturday morning meeting, with other PMs who were either managing problem programs or programs of special interest. We were "explaining" our program status to an audience which included senior Department of Defense Acquisition officials, a room full of U.S. Air Force generals and colonels, and my "senior management," Boeing's chief operating officer and the president of the Defense Business Unit. My job here was to explain the issues with the program, the root causes, proposed corrective actions, and the path forward. As you will appreciate, having been on the program such a short time, I was very nervous but I worked hard to be prepared and did my best to anticipate questions.

The audience was not demonstrably hostile; call it, guardedly cordial . . . albeit, a bit skeptical. It was clear that they would hold me to my promises, to get the program back on plan. If at any point I was failing, even to some small degree, I could expect an invitation to return for a less-friendly grilling.

This program needed a lot of change: in culture, organizational structure, people, and process discipline. Solutions? Read on . . .

1. I built relationships with my team, customers and suppliers, all of whom were new to me. I spent time at their locations to get to know all the stakeholders. I listened intently to gain an understanding of their perspectives on the program, what was working, not working and hear their suggestions for improvement.

2. Everyone on the team was encouraged to let me know what was needed—and to not try to hide or soften "bad news." Bad news doesn't get any better with time.

3. We held daily meetings to discuss tactical priorities for the day and the week, align work tasks, ensure that needed assets were in place, and share supplier and customer feedback (a vital component of any program). We established benchmarks to track progress, to ensure that the "must-fix" items were being completed on time and with technical rigor. It didn't take me long to realize that some members of our team were not quite prepared for some critical milestones. There were, for example, timing issues in the release of data and drawings to sup-

pliers, who would be fabricating some elements of the system . . . elements which we needed in-place for assembly at a point-certain.

Utilizing the simulators with the customer, our program test pilot was able to identify shortfalls in the design of the new digital cockpit. Truth to tell (as with the *Harrier* wing), things got worse before they got better.

Our estimate, that we could reuse about 75 percent of the Israeli-developed software, was dramatically dropping; it would be at about 20 percent. As we learned, that software was not consistent with U.S. Air Force practices, especially training student pilots on Basic Flight Maneuvering, Communications, and Navigation. Further, the program was too optimistic on the use of COTS and NDI hardware—it would not meet the system level requirement, which added to development time and cost.

The team worked hard to address the must-fix items and put an aircraft into a Development Test, conducted by a Combined Test Force (CTF) made up of pilots from the U.S. Air Force and Boeing. Logisticians were on hand to help with maintenance questions. The CTF provided an opportunity for the customer and Boeing pilots to fly together and collectively determine if the aircraft was meeting its design requirements. The aircraft was configured with wiring unique to flight testing to collect relevant data from the flight for both validating analysis and the purposes of root cause analysis should problems arise. In summary, this Development Test (DT), which included operational testers (OT) in a combined DT/OT process, was considered successful and the aircraft was ready for Operational Test.

Operational Test is a huge hurdle for programs reaching that stage. The aircraft is turned over to an independent group of testers who evaluate the aircraft for operational effectiveness and suitability/ reliability. The flight test wiring is removed to provide an operationally configured jet and the maintainers use our technical data for troubleshooting and maintenance, and it needs to stand on its own as Boeing pilots, engineers and logisticians are not included in this evaluation. With a challenging KPP for system level reliability, that would be a big focus area for the testers.

Despite our best efforts to work with our industry team, when we entered operational test, we were having on average about one maintenance action per flight. This translated into one-hour MTBF, against the requirement of 60 hours MTBF. There was obviously a significant gap between requirement and performance. While system level reliability generally increases over time, we were experiencing too many maintenance actions and equipment failures. The customer called a pause to operational test.

Progress payments were suspended and monthly meetings at the Pentagon started shortly thereafter.

It was along about here that I saw the movie, *Apollo 13*, about a failed 1970 space mission and the efforts made to save the crew. I thought if NASA could do what they did, we could fix these issues.

If you are not familiar with the Apollo 13, the spacecraft was in its third day of the mission when a routine procedure resulted in an oxygen tank explosion, causing NASA to abort the mission to the moon.[4] The explosion rendered the command module unsuitable for sustaining the three-man crew until re-entry to earth. The crew used the Lunar Module as a "lifeboat" to get them to a position of reentry. There were multiple technical challenges that had to be overcome. What impressed me so much is how Gene Krantz, the Flight Director, led the team consisting of NASA and the contractors, working together with the astronauts, anticipated and overcame every technical challenge, to ensure a safe return for the crew. The collective team made use of prototyping to devise a way to filter out dangerously high levels of carbon dioxide from the lunar module—using materials onboard the spacecraft. In addition, an astronaut at NASA's facility used a simulator to develop a procedure to power up the damaged command module for reentry into the earth's atmosphere. The Flight Director insisted on a written procedure for the crew so there would be no ambiguity in what they had to do. This team exemplified leadership and technical excellence.

Through a methodical "analyze, fix, and test" approach, the Boeing/Industry team was determined to put the T-38C Avionics Upgrade Program on a path to success.

Key actions:

1. I asked for help—and got it. We had a new supplier providing the Multi-Function and Electronic Engine Displays. We brought on one of our best display engineers, who would help accelerate the development process. I was able to add a deputy PM with strong leadership and mission avionics expertise, and an awesome hardware-software integration engineer. I assigned him as chief engineer; he was in place the day after the operational test was put "on hold."

2. We brought in a strong reliability engineer from the C-17 *Globemaster* III program, who conducted an

independent review. She taught us how to determine and depict a realistic growth curve. The MTBF Key

Performance Parameter (KPP) value would not be realized the first day, week, month or even year. It wasn't until we had accumulated thousands of flight hours did we demonstrate the KPP. This engineer also helped us with explaining the rationale behind the growth curve to the customer.

3. With facts and data from operational test conducted before the pause, the team was able to address root causes associated with hardware failures and drive the right corrective actions.

4. At the end of our daily team meeting, we played the audio clip from *Apollo 13* of actor Ed Harris' portrayal of Gene Kranz's famous line, "Failure is not an option." Perhaps a bit corny, but it wasn't a complete day without hearing it.

5. When we resumed operational test, we utilized the first three Low Rate Initial Production (LRIP) aircraft instead of the two Engineering, Manufacturing and Development (EMD) aircraft. Since operational testers require production representative hardware, we had three pristine jets for them to fly. With these new aircraft, which incorporated a series of corrective actions, we would begin Initial Operational Test and Evaluation (IOT&E). Every day, I would ask the team, "How many "Code 1" flights did we have today? Code 1 flights were completed test flights without a maintenance action requested from the aircrews. Our team knew the outcome for each day of flight, and the running tally for the test.

Finally, it was time to meet with the Air Force acquisition leaders, and I knew we were pretty close . . . but . . . was "close" close enough? I will never forget sitting in the conference room, with the testers on video teleconference, when the question was asked, how did the test program go? The answer was "Operationally effective and potentially suitable." These five words would result in a successful Full Rate Production decision and a resumption of progress payments. I was so happy for our team and that we were ready to provide the customer with the capability they needed and deserved.

Our modification center at Williams Gateway in Mesa, AZ, was fantastic. We brought in airplanes, the center performed the modifications, and then customer pilots would fly them back to one of six Training Command bases. Soon after we started full rate production, with things under control, my deputy program manager would take over as program manager.

Over the course of the T-38C modification program, which lasted about seven years, the team would introduce a number of significant improvements. Many were based on Lean+ concepts which over time brought the direct-labor cost down to 25 percent of that of the first unit and significantly reduced cycle time.

One "Lean" improvement was to transition the factory floor from "stationary" assembly, where the aircraft remained in one spot perhaps for several weeks as teams came up one after another, to a "moving line" assembly where the aircraft is shifted or "pulsed" every few days to a new "work station."

With Lean manufacturing, parts, tools, and instructions are kitted in an organized fashion and brought ship side, so mechan-

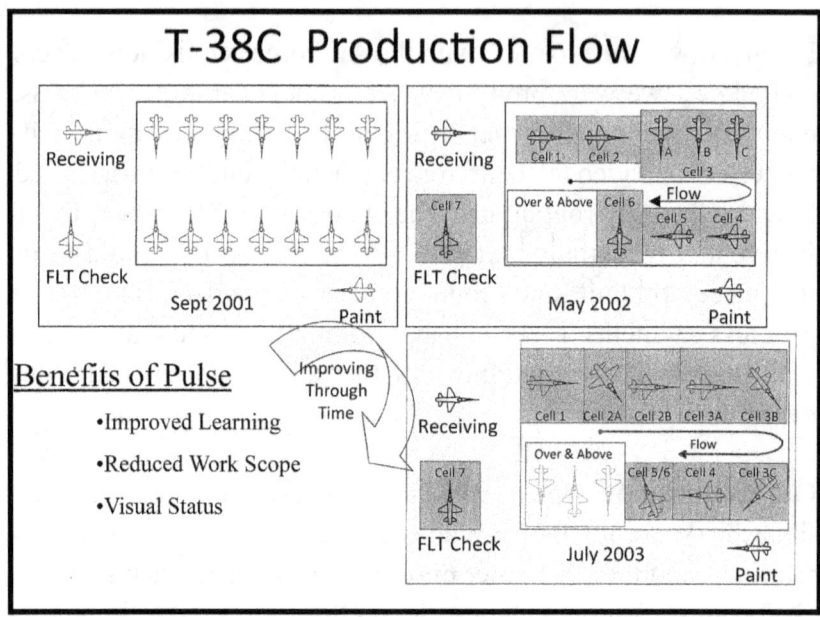

Figure 10.6 T-38 Lean Transformation Factory Floor Layout Evolution (Copyright, Boeing)

ics or electricians have everything they need, and will know how much work is to be completed each day. "Status" is posted on the

factory floor and for support teams, updated frequently, so everyone will know the "plan for the day," follow progress and be made aware of issues.

Improving the rating of "potentially suitable" was a running challenge, to move up the MTBF curve. With the agreement of the customer, we dedicated several months to test, measure, and demonstrate the MTBF improvement. And, we reached the 60-Hour MTBF requirement.

We modified more than 460 aircraft for AETC and completed the program. As of this writing, the fleet of T-38C *Talon* aircraft (and supporting flight simulators and courseware) continues to be the U.S. Air Force's primary trainer for fixed-wing pilots. This system is programmed to be operational until, at least, 2025.

Key Case Study Takeaways:

- Designing for reliability and supportability starts at the very beginning. Trying to fix low reliability once aircraft are fielded is nearly impossible and cost prohibitive; it has to be designed-in.

- Good leaders attract great people; it makes all the difference between success and failure.

- The programs that ask for help are the most successful.

- For any program with an MTBF requirement, I highly recommend a "demonstration phase."

- One of the company's Lean experts said, "Engaging the hearts and the minds of the work force is absolutely essential to a transformation success. When the work cell teams own their performance, are responsible for continuous improvement, and meeting cost and quality performance, improvement is endless."

Case Study: United Kingdom Chinook Programs

Ensuring Readiness Everyday for the United Kingdom's Royal Air Force Chinook Fleet

Boeing—the OEM—won a Through Life Customer Support[5] (TLCS) performance-based logistics (PBL) contract in January 2006. We were to ensure that 27 United Kingdom (U.K.) *Chinook* Mk2/2a aircraft—out of a 40-aircraft fleet—were mission-ready, every day.

We provided in-country program management, engineering technical support, and supply chain management for spares and repairs. We subcontracted heavy maintenance—initially, through an U.K. government facility (which later was privatized). Also, we signed with two RAF bases for specific repair capabilities.

To meet the "operational ready" target, we would have to complete depot maintenance, for each aircraft, on-time. Here, we saw an opportunity to implement factory floor "Lean" processes, and brought in experienced team members; to lead this effort, we gave a long-term on-site assignment to one of the best. The team reduced depot-level maintenance cycle time and met all goals. Another challenge: to provide timely engineering support for aircraft requiring repair. For this, we would have to "reach-back" to the original design authority, the Philadelphia *Chinook* Engineering team, which provided safety analysis and airworthiness certification. (I should note, *Chinook* Engineering was not just a support "archive" for our effort, but continued to develop and build new models for the U.S. Army and variants of those models for international customers.)

The Boeing *Chinook* Mk2/2a models—the original program dated back to 1961—were long out of production and therefore the responsibility of Global Services & Support. The engineering organization was managed and led by a director with several product teams to provide whatever tasks were required through program-aligned chief engineers—who had a dual responsibility to program managers for the U.S. Army, United Kingdom, Netherlands, Canada, and Italy programs, and for other international customers. To further add complexity, the *Chinook* program was at the same time a viable production program, and the engineering group reported in to Boeing Military Aircraft. This was, however, an excellent example of leveraging the best resources and capabilities within Boeing.

This was the first international PBL program. While we staffed the TLCS team with solid engineering talent, many didn't have *Chinook*-specific experience and it would take time to transition the more complex engineering work. With the center of engineering expertise in Philadelphia about 3500 miles away from the program team in the U.K., along with the five-hour time difference—not to mention, an engineering team that was not part of my organization, "building relationships" was vital. When you don't have direct control over a team on which you must rely for success, what can you do?

- Hold onsite introductory meetings in Philadelphia to meet the *Chinook* program chief engineer and key IPT leadership.

- Communicate TLCS program objectives and success criteria throughout the team.

- Identify a senior engineering manager from the *Chinook* Integrated Product Team who would be aligned with the in-country chief engineer for coordination, establishing IPT priorities and help needed.

- Establish clear responsibilities—which organization had the lead on specific engineering tasks, with a goal of eventually transitioning much more work content to the in-country program team.

- Engage frequently with the in-country chief engineer and the U.K. *Chinook* IPT leader in Philadelphia to review performance metrics and leading indicators to highlight potential issues in the future and assess help needed.

● Communicate your expectation of the need for pro-active support during the depot level maintenance process and support the Lean transformation in the factory.

After start of the TLCS Program, Global Services & Support hired an in-country executive for all Rotorcraft Support, a man with extensive customer knowledge and experience of doing business in the U.K. His goal: to grow the Boeing presence for Rotorcraft Support and be the provider of choice to the U.K. Ministry of Defence. U.K talent with needed skillsets would be phased in, to put more of the direct execution in-country.

After a slow start, the TLCS program was solid. The customer increased the flight hours on TLCS, and the team rose to the challenge while meeting the "27 aircraft-ready each morning" KPP. Under the performance-based contract, the increased flight hours enabled the program to bring in higher than planned revenues.

There was another opportunity for Boeing to help the U.K. customer. There were eight Mk3 model *Chinooks* developed with a digital cockpit and built some years prior, which failed to achieve airworthiness certification. They were, therefore, in storage and not flying. Now, the customer contracted with Boeing to "revert" them to an analog cockpit (similar to the Mk2/ 2a fleet), referred to as the Mk3 Reversion (Mk3R) Program.[6]

The reversion work was conducted by an in-country supplier at an RAF base in southern England, and with excellent on-site leadership, the team would work through both engineering and operations challenges. An added challenge: the Mk3R *Chinooks* had larger fuel tanks than the Mk2/2a models. Once the team began flight-testing, the Mk3R aircraft exhibited some pilot handling qualities in certain maneuvers that were different from the Mk2/2a fleet. Therefore, working together with the customer, we established a pilot training sequence that covered the differences.

Once this program was completed, the Mk4 Program was established to develop a new digital cockpit using an in-country supplier of mission avionics.[6] The new cockpit was to be installed in all Mk2/2a and M3R aircraft. Our *Chinook* IPT would lead the engineering. The eight Mk3R aircraft when upgraded with the Mk4 cockpit would be designated as the Mk5 aircraft. Once complete, the plan was for the fleet of 46 aircraft to be maintained and supported through the TLCS program.[7] There would be many face-to-face meetings with the *Chinook* IPT in Philadelphia and with the customer and supplier in the U.K., to review program status, ensure prioritization for resources, help address technical issues, and engage with both customers and suppliers.

As the development was primarily avionics and software, we built a recovery plan that would provide incremental capability to our customer. With this approach, flight-test of the aircraft could begin. We assigned Boeing engineers to work onsite at the supplier's facility to ensure software and avionics integration were being well-executed. Further, we ensured prioritization of resources, both within Boeing and with our supplier. The Mk4 program was granted initial release to service May 2012, and declared an initial operating capability in June 2012.

The digital cockpit provides the U.K. customer with an integrated solution, which reduces pilot workload through a more effective human machine interface (HMI) that the analog cockpit did not have. As of this writing, the TLCS Program is in process of upgrading all their Mk2/2a and Mk.3R *Chinook* helicopters with the Mk4 digital cockpit. The program continues to provide the customer with the required capabilities and readiness levels.

Key Case Study Takeaways:

• The TLCS Team is ensuring readiness for the U.K. RAF's *Chinook* Fleet every day.

• To be successful in the international market place, have presence in that market and leadership that knows and understands the customer. The U.K. *Chinook* Program is an excellent example of what can happen with the right leadership and team in place.

• When programs successfully execute on existing business, new business opportunities follow.

Case Study: A-10 Re-Wing Program

Re-Winging the A-10 "The Warthog," the United States Air Force's Premier Close-Air Support Aircraft

When aircraft are designed, the contract generally requires that the basic airframe will be good for approximately 6000 flight hours. Of course, military aircraft (especially) are not always flown according to initial design and intentions. Thus, programmed service life may be compromised. With the A-10, the wings became a particular issue, and "improved" wings were being added to the inventory . . . but not enough, and not soon enough to support the missions being flown.

In 2007, Boeing won a "Build-To-Print Plus" contract that included a development program and new production wings for the A-10 *Warthog*.

- "Build-To-Print" means a contractor builds product to a specified set of engineering drawings, process specifications, and work instructions.

- "Build-To-Print Plus" adds some "design for manufacturing" considerations "Plus" design changes to improve structural life, but the team would be limited on what DFMA improvements could be incorporated per customer requirements.

- A key Build-To-Print Plus limit on this contract: we would not perform full scale structural testing on the new wing. This meant that primary load paths, and wing to fuselage attachments and backup structure, had to remain the same to every last detail. Further, it meant that critical design features as small as "fastener hole diameter tolerances" would need to remain the same to not impact "internal load distribution of the structure under flight maneuvering loads." Wing bending and "torsional stiffness" had to be within a specific margin as to not impact other critical flight characteristics.

We were required to make "fatigue improvements" based on fleet history, and "design for supportability" changes would be made to facilitate maintenance. As with the T-38, the A-10 was not designed or built by Boeing. We obtained the 1970s-era drawings from manufacturer Republic Fairchild and per customer request we converted them to 3D master models. This was a key step in ensuring first time quality engineering. There were some issues discovered in the original design that needed to be decon-

Figure 10.7: A-10 Aircraft with new wing—Nov 2011 (Copyright, Boeing)

flicted prior to starting the product definition (that is, design of the wings) process. Anomalies found were identified and corrected during the drawing-to-3D conversion process. Given that the 3D models would be the basis for every aspect of creating new engineering for the wing, the team took the time to make sure it was right before we began the systems engineering and design. Along with the engineering drawings came the Republic Fairchild "process" specifications—with a minor complication, they did not match anything we were accustomed to. This would be an issue, later, when we actually started building wings. An added challenge was an incomplete set of "master tooling" required to ensure new wings would fit on old fuselages.

We put in place a terrific team to perform the product definition. The engineering team would work with other functions such as, supplier management and procurement, operations, quality, and product support—and our partner, the U.S. Air Force. The engineering would take place in St. Louis and the build would take place in our Macon, GA manufacturing facility.

The program was organized around four Integrated Product Teams, —Center Wing, Outer Wing, Control Surfaces, and Subsystems (fuels, electrical, hydraulics). The teams benefitted from the latest generation of computer-aided design and analysis tools—all stakeholders could literally "see" the structure, mechanical and electrical systems. Weekly technical reviews addressed staffing issues, performance to plan, technical risks and issues, and would provide help to the team, as appropriate.

One of our IPT's would run into trouble, and get behind. They missed some benchmarks (or, in engineering-speak, "leading indicators of completed drawings to plan") which showed we would not complete the work on time. This would impact procurement of parts and start of assembly. Too much work was in process and not enough was being finished. We shifted to a "focus and finish approach," essentially, dealing with one task or issue at a time. The team was able to recover to plan.

The team implemented some innovative tooling concepts. Instead of the traditional assembly jig—often, a large structure which helped position parts but which over-time, through wear and tear, would lose precision—we started using some major parts themselves as "tools." Properly "prepared" in the machining phase, they easily substituted for a jig.

As noted above, the contract called for improvements in "fatigue" and "supportability." To address the first, we utilized upgraded aluminum materials, strengthened some areas, and re-designed parts which had been failing too early. For supportability—one simple example, by increasing the diameter of access holes by 1" and by modifying the attachment for some door-covers, we provided easier access for the maintainers. Virtual Reality (VR) reviews provided collaboration and feedback, not just from the design product teams, but also from operations . . . hundreds of miles away.

Once we completed the engineering process and had authorized fabrication of parts, we discovered we were not yet home-free. While we were a proficient manufacturing facility with a strong track record, building product to another company's legacy design was challenging.

- We over-estimated probable production rates and underestimated complexity on the initial four A-10 wings.

- There was a high level of rework, repair and scrap during manual hole drilling operations on the first few wings.

- Because of the number of non-conforming conditions on the early wings, we had to convene Material Review Boards (MRB) with program and government engineers, who worked together to review how to fix, or in engineering speak return the "as-built" to the "as-designed" configuration and recommend corrective action. This significantly added cycle time to the build process.

- A drawing-based design is imperfect in defining the

exact manufacturing requirements, unlike model-based definition which integrates the design and build into a product. A key lesson is, to truly "build to print," you need more than drawings or engineering—you need an understanding of how the original was built.

There was a significant level of intervention by Operations, Quality and Engineering leadership to augment the work force, adjust factory floor instructions, and provide additional training. The team worked through the build process challenges but not until there was a several-month delay in delivery of the first wings. Once various problems had been addressed, the program performed well.

Key Case Study Takeaways:

• Like the T-38, the A-10 is a great example of executing an upgrade to an out-of-production aircraft for which Boeing had not been the OEM.

• Customers depend on contractors to improve supportability and ease maintenance. You must incorporate these factors from the beginning; you can't wait until product is fielded to address these considerations.

Case Study: Gerald R. Ford *Class Aircraft Carriers*

Designing and Building the Newest Generation of Aircraft Carriers That Will Provide Extraordinary Capability and Global Reach

In recent years, military aircraft programs have suffered cost overruns and schedule slips in development and demonstration, due, in part, to efforts to incorporate technologies that were not as mature as expected. This Case Study will show that "Military Aircraft" is not the only category of DoD programs now being challenged through the Product Life Cycle process of Development and Demonstration.

The *Gerald R. Ford* carrier class will replace the USS *Enterprise* and the *Nimitz* class aircraft carriers. Having been on a *Nimitz* class carrier, the *John C. Stennis*, CVN 74, I am awed at the capability of today's fleet of nuclear aircraft carriers and have tremendous respect for the contractors and people that design and build these impressive war-fighting ships.

The launch date for the lead carrier, USS *Gerald R. Ford*, CVN 78, is scheduled for August of 2016 and construction for the USS *John F. Kennedy*, CVN 79 is underway. Some of the capabilities and technologies of this new class of aircraft carriers:[8]

- They will carry up to 90 aircraft, including the F-35 Joint Strike Fighter, F/A-18E/F *Super Hornet*, E-2D Advanced *Hawkeye*, EA-18G *Growler* electronic attack aircraft, MH-60R/S helicopters, along with a variety of unmanned air vehicles.

- They are planned for a sortie rate at 160 a day (compared with 140 a day for the *Nimitz* Class) with surges to as high as 220 sorties a day. This has led to design changes in the flight deck. The deck-edge elevators are reduced from four to three, aircraft parking space is enlarged, the "island" (the navigation and command center) is smaller and re-located.

- A new electromagnetic aircraft launch system (EMALS), replaces the traditional steam catapults. EMALS offers finer acceleration control, which leads to lower stresses on the aircraft and pilots, provides a slower launch speed for unmanned air vehicles, and allows a wider window of wind-over-deck speed (required for the launch sequence).

- An advanced turbo-electric arrestor gear, which has an electro-magnetic motor, is able to reduce the maximum tension in the arrestor cable and thereby reduce the peak load on the arrestor hook and, in turn, on the aircraft fuselage.

- Advanced weapons elevator systems and power generations systems will support mission readiness and operational tempo objectives.

As of this writing, the lead carrier, CVN 78 is more than $2 billion over cost, incorporation of some new technologies have been deferred, and the launch date will likely slip. A Government Accountability Office report as far back as 2007 warned that costs were underestimated and critical technologies were not at the desired maturity level.

Paraphrasing, the program is in construction while still working on development. "The carrier's problems are by no means unique; rather, they are quite typical of weapon systems," GAO said.[9] The root causes include over-optimism, and pressure to over-promise. These are not unique to aircraft carriers. New technologies should be at a defined maturity level when preliminary design begins, not after "development" is finished. The consequences are, well, obvious. As an engineering leader, it was my job to ensure new technologies were at the desired design and integration level of maturity. More details, see Chapter Eleven.

For reference, basic categories of "new technology" for military aircraft include (but are not limited to) sensors, digital avionics, battlefield interoperability, new material systems, and production processes.

New technologies, of course, are vital and development usually comes from contractor-funded research and development (R&D) or government-funded R&D. "R&D" implies successful evaluation, analysis, laboratory testing, and prototype demonstration before any contracts are awarded. At times, as we have seen, awards of contracts based on new technology may be, well, premature. At my end of the business, it falls to the engineers—who were not necessarily involved in contract negotiations—to assess, integrate, and monitor to the best of their ability. ∎

Key Case Study Takeaways

• One of the biggest drivers of cost overruns and schedule delays: failure to achieve the desired level of technological maturity before entering production, which drives rework and may contribute to a mis-match between development, test, and production. Mature technology leads to reliable, safe, and mission ready products.

• Applying lessons learned from both industries, (aircraft and ship building) can make a difference in meeting the needs of the warfighters, on cost and on schedule.

---

1. Boeing 737 AEW&C Wedgetail Early Warning Aircraft, Australia, http://www. airforce-technology.com/projects/737aewc/ (Retrieved 22 NOV 2015)

2 This concept is important for engineers to understand. While hard work is truly appreciated, it is the results that count. People get measured on results, not effort.

3. Base Realignment and Closure (BRAC) https://en.wikipedia.org/wiki/Base Realignment and Closure (Retrieved 21 SEP 2015)

4. Apollo 13 NASA, 8 July 2009 https://www.nasa.gov/mission pages/apollo/missions/apollo13.html#.Vkyb_N5CjjA (Retrieved 18 NOV 2015)

5. *Chinook* and *Chinook* Through Life Customer Support http://www.boeing.co.uk/products-services/boeing-defence-uk/aircraft-services-support/chinook-tlcs.page? (Retrieved 12 NOV 2015)

6. Elite UK Forces http://www.eliteukforces.info/air-support/7-Squadron/chinook.php (Retrieved 12 NOV 2015)

7. Two Mk2/2a aircraft out of the original 40 aircraft were lost after the initial contract award. Therefore the fleet size of *Chinook* aircraft for digital cockpit modification was 46.

8. *Gerald R Ford* Class (CVN 78/79)—US Navy CVN 21 Future Carrier Programme, United States of America http://www.naval-technology.com/projects/cvn-21/ (Retrieved 12 NOV 2015

9. Fabey, Michael, *Ford Based On Unrealistic Business Case,* GAO Says p4. *Aviation Week's* Daily and Defense Report, 05 OCT 2015

CHAPTER ELEVEN
OPERATIONALIZING TECHNICAL EXCELLENCE

Perfection is not attainable, but if we chase
perfection we can catch excellence.

Vince Lombardi
NFL Football Coach

W hat is technical excellence? How is it achieved? Why is the chief engineer the guardian of technical excellence? Technical excellence is about "Supporting the Business and Protecting the Enterprise." My approach to operationalize technical excellence was the following:

Support the business:

- It starts with First Time Quality (FTQ), which means completing the job right the first time, and without passing technical risk downstream to test or production. In complex military aircraft programs that is a tall order, but necessary to remain competitive. We have to strive for continuous, if not step-function, improvement.

Steps to achieve first time quality:

- Staff your programs and integrated product teams

with the right people in the right place at the right time.

• Identify and strengthen core competencies consistent with the organization's business model.

• Implement a Lean+ "Focus and Finish" approach to prioritize tasks, to not starting tasks before all the data to finish is available, all the while minimizing work-in-process. Hold monthly reviews with each of your programs to share stories of success.

• Assign technical lead engineers within Integrated Product Teams to "advise, assist and check," especially for the more junior team members.

• Utilize peer reviews to double check the more complex analysis and modeling tasks.

• Establish proactive and collaborative, working together, relationships with program IPTs and suppliers.

• Implement a robust supplier technical engagement and oversight process.

• Engage with teams at and between critical points in the development process, using a "Trust But Verify" approach.

• Leverage the power of engineering talent across the organization to address specific complex issues.

• Utilize Engineering Centers of Expertise for specialty engineering tasks.

• Perform a Root Cause and Corrective Action (RCCA) to address technical problems.

• Help position the business for competitive differentiation through operational innovation. This includes the implementation of step function improvements in processes and tools, technologies, rapid prototyping, and focused R&D spending.

Protect the Enterprise

- Establish and implement an "Engineering System."

- Make technical decisions in the best interest of the enterprise.

- Identify technical risks before they become issues and work with the program engineering team to mitigate.

- If risks materialize into issues, help the teams work through problem resolution, which include anything from staying out of their way to actively engaging to solve the problem, as appropriate. Either way, ensure that a disciplined approach to Root Cause and Corrective Action is taken.

- Effectively communicate risks and issues with business, program, and functional leadership to provide situational awareness and avoid surprises. Provide go-forward action plan and identify help needed.

- Disposition readiness to proceed on from one phase or "Gate" of development to another; for example, moving from a firm configuration phase (which defines requirements, architectures, and interfaces) to the next phase, preliminary design. I highly recommend: do not conduct a Gate Review if anticipated action plans or outstanding work can't be closed within 30 to 45 days.

- Approve programs to proceed into ground and flight

test.

● Make decisions to ground aircraft due to real-time facts and data, problems or incidents, or safety concerns.

Indeed. My job was to "Support the Business and Protect the Enterprise." I applied this practice to every chief engineer assignment I had in The Boeing Company. When necessary—and generally speaking it was the exception—I would not approve a Gate Review, not approve a flight test, and would not hesitate to ground aircraft if the facts and data warranted. Sometimes people disagreed and said I was too conservative, but the buck stopped with me.

Creating the Right Culture and Work Environment

During development of the F/A-18, McDonnell Douglas established a set of Program Management Best Practices, one of which was "Help Needed," an encouragement to come forward and ask for help, any kind of help. Typical examples? Asking for assignment of new staffers with specific skill sets, addressing facility space issues, or eliminating non-value added administrative tasks. "Help Needed" was usually the last chart in every team's weekly briefing to Program Management. If no help was needed at that time, the slide said "None."

The goal was to create a culture where not only was it okay to ask for help, but ensure that requests were acted on. This practice would expand across all of McDonnell Douglas Aerospace, and be in place when McDonnell Douglas and Boeing became one company. It didn't take too long to see that the programs or teams that asked for help were more successful than those that didn't. The F/A-18 program has been successful throughout its evolution, from the 1970s A/B models through the E/F *Super Hornet* and on to the EA-18G—a variant of the E/F which performs electronic warfare missions.

In 2002, as a companion to "Help Needed," the Military Aerospace Support organization instituted a culture of "Don't Shoot the Messenger" and "It was Okay to bring bad news forward" as two of tis operating principles. The President of Aerospace Support had been vice president and general manager of the C-17 *Globemaster III* program, which had significant early issues—to a point that the U.S. Air Force announced it would cut off procurement at 40 aircraft. But—adding a few tricks from the Malcolm Baldrige playbook[1] (especially, bringing in critiques from outside the industry)—he turned things around and the C-17 program grew to more than 200 aircraft for the U.S. and international customers.

I set the same expectations for program chief engineers on my team and held them accountable.. And I maintained personal engagement, adding a "Trust but Verify" dimension. President Ronald Reagan used this philosophy when dealing with the Soviet Union during his presidency.

The complexity of aerospace programs in any phase, but especially when in development, requires an open culture . . . where people can come forward and ask for help. I can't emphasize enough the importance of transparency and working together collaboratively.

The Engineering System

We all know that people will make mistakes, but the system can't allow mistakes, especially traveling from one program phase to another. Meaning, we need a system of checks and balances that is robust enough to find errors and take appropriate action to fix them before they can impact the next step in the program development. In past assignments, I saw team leaders who attempted to keep everything in the room, so to speak, whether from pride or ignorance—when there was a vast store of expertise and experience just outside the door, waiting to be tapped. It was clear to me that we should leverage strengths from across the company, as necessary, creating a wide-ranging Engineering System.

The Engineering System I used in Boeing Military Aircraft

The Team:

• BMA engineering leadership, led by myself—with Responsibility, Authority and Accountability (RAA) for technical excellence across the business.

• Boeing Defense, Space and Security (BDS) senior chief engineers and functional directors—their RAA included:

  ○ Providing technical guidance on complex issues.

  ○ Processes and tool development and maturity.

  ○ Enterprise skills and staffing.

• Program chief engineers—technical excellence across the program.

• Integrated Product Team working with leadership of the Engineering Centers of Expertise.

- Technical lead engineering—advise, assist and check work of the engineers assigned to their work group.

- "Technical Fellowship" (engineers selected through a rigorous process identifying them as corporate and industry leaders in their fields of expertise)—specialty engineering skills

- Skilled program engineers—execute program work.

The "Concept of Operations" or, How it Operated

Bi-weekly, my leadership team and the senior executives across BDS, part of this Engineering System, would meet and discuss known technical issues and explore opportunities to get out in front of them. The primary focus was on the highest priority programs and strategic initiatives, but we kept an eye on programs and projects which at the time were not at the top of the list.

We worked from a spreadsheet—risks, issues, and opportunities—to track specific program items, looking for unresolved technical, schedule, and/or cost implications. We updated the database weekly, which gave us a foundation for steps to be taken, perhaps providing additional help or engaging directly to better assess the problem. Our collective relationship exemplified teamwork in the best interest of the programs and the business. Collaborative discussions between the senior leaders, the program chief engineers, and the teams were the most direct and efficient way to determine the way forward. The Engineering System was very effective when the stakeholders were "all-in." With any failure, especially one that resulted in rework, the root cause could usually be attributed to a breakdown in the interfaces between team members within the system.

Deep Dives

I adopted the "Deep Dive" collaborative brainstorming technique to stay abreast of technical performance, technical risks and issues, schedule and cost impacts, and gain customer perspectives. This was a process that was not unique to The Boeing Company, but was widely used throughout industry. Former General Electric CEO and Chairman, Jack Welch wrote about his approach toward "deep dives" in his book *Jack—Straight from the Gut.* "It's spotting a challenge where you think you can make a difference—one that looks like it would be fun—and then throwing the weight of your position behind it. Some might justifiably call it meddling."[2] Another executive—Sean Connolly, CEO of ConAgra—wrote that an outsider joining ConAgra as a senior executive would lead a deep dive "to search for ways to increase efficiency in the manufacturing, procurement, transportation, warehousing and other functions for its retail brands."[3]

Depending on the phase of the program under review, I would hold deep dive sessions monthly, more often if necessary. Attendance might include senior leaders from supplier management and procurement, operations and quality and the operating executive. If a highly complex technical challenge was being debated, some of my engineering peers might be included. I would travel to the sites where work was being done; whenever possible, these engagements would take place at the location where the work was being performed.

My deep dives were informal, but we always had an agenda. Rules of engagement included (1) don't make any special charts for this meeting (I wanted the teams to use existing material), (2) I would take more action items than I gave, and (3) it was to be a dialogue, not "death by viewgraph." We might include "virtual reality" reviews of a design, and, if practicable, take a walk through the laboratories, the factory, and in or around the aircraft. Rarely did a review last more than four hours. My objective was to make this a value-added half-day, to generate critical thinking amongst us all and give me a perspective on the maturity of the engineering. Whenever possible, I always welcomed an opportunity to have a roundtable discussion with teams and emerging leaders in conjunction with technical engagements.

Technical Deep Dive meetings usually followed the following agenda:

1. The program chief engineer would kick the meeting off and provide an overview of the team's performance, the technical health of the program using a comprehensive set of metrics, discuss the plan forward, and highlight technical risks and issues the team was facing. When possible, we would discuss customer feedback received since the last deep dive.

2. The IPT leaders would follow and share similar information as it pertained to their team.

3. My immediate participation?

a. I wanted to hear, what was the feedback from customers?

b. I would seek to assess the level of understanding behind the data.

c. I looked for a balance of leading and lagging indicators of performance, including quality and productivity/throughput metrics. If teams were behind plan, I wanted to know about the recovery plan; and would assess if there was sufficient detail that would enable closure.

d. If teams were contemplating changing direction or evaluating potential solutions, I wanted to know if second and third order consequences had been considered, and if not . . . I would "suggest" that they do so.

e. I would evaluate systems-level thinking and understanding and focus questions on integration between teams and the maturity of physical, functional and logical interfaces. If issues arise, these are generally the areas where they will appear.

f. Also, I would look at ground rules and assumptions; specifically to see if they were still valid and if we had anchored assumptions with analysis or test data. For example, if a team estimated a specific level of software reuse, were they performing to plan? If they were off plan by any measurable amount, that would indicate some level of cost overrun and schedule delay.

g. I wanted to see evidence of Lean+ deployment, with key metrics that measured process health.

h. Depending the program scope, I would inquire, "How was the team designing for ergonomics, manufacturing, and supportability?"

i. I would review technical risks and mitigation plans—I wanted to know if the team had a high degree of confidence in their plan to mitigate the risk, or if there was a plan B, standing by. Depending on the answers, this would be an area for further review.

j. Assess if the team was accomplishing the amount

of work needed each day to meet milestones and budget commitments—also known as "Value of a Day."

k. The wrap-up and discussion would include action items, mostly assigned to me. I made it a point to deliver on promises.

Workplace Safety

When it comes to protecting the enterprise, I can think of no more relevant topic than workplace safety. The following, from a United States Labor Department article on the "Business Case for Workplace Safety," provides context.

> Workplace fatalities, injuries, and illnesses cost the country billions of dollars every year. In its 2012 Workplace Safety Index, Liberty Mutual estimated that employers paid almost $1 billion per week for direct workers' compensation costs for the most disabling workplace injuries and illnesses in 2010. Employers that implement effective safety and health management systems may expect to significantly reduce injuries and illnesses and reduce the costs associated with these injuries and illnesses, including workers' compensation payments, medical expenses, and lost productivity. In addition, employers often find that process and other changes made to improve workplace safety and health may result in significant improvements to their organization's productivity and profitability.[4]

One billion dollars a week nationwide, that's a staggering number. Can you imagine how many people are not working because of a workplace injury? Not to mention short-term suffering or long-term disability. Now, some—perhaps most—workplace-safety issues are not directly related to "engineering" but revolve around slips and falls and stupidity (such as, reaching into an operating machine to retrieve a dropped screwdriver). The relevant question, here: what is the role of engineering in addressing safety? The answer: design with "safety" in mind.

As I recall, during my days as program chief engineer on the F-15 program, there were problems connected with drilling through titanium, which is a very hard material. We were seeing worker fatigue, broken drill bits and injuries to the hands, not to mention out-of-conformity holes leading to scrap or rework. We came to realize that we had to design ahead, and brought in the Manufacturing Methods Group for assistance. Today, the F-15 program uses special drill motors which can be locked in place (no more operator struggling to keep on-point); that allows the operator to easily control the speed (revolutions per minute) and the feed (inches per minute) of drilling, and coolants are used to keep temperature (of both the drill bit and the hole) within a safe range. Supervisors on the floor have spec sheets they consult, to ensure compliance. Further, the addition of an automated contour drilling machine for the aft fuselage structure has significantly improved hole quality and reduced safety related metrics.

Another example: have you provided enough space in, or access to, a structure (as in, a fuselage or a wing), to accommodate a mechanic trying to connect a wire or install a part in such a manner that it eliminates—or at the least, reduces—acute discomfort or pain which, over-time, could lead to disability? As we evolved our product definition processes, we increasingly considered ergonomics in the designs. Engineers today, for example, model a mechanic installing a part in a tool, or a piece of equipment in an aircraft to see how easy or difficult it may be. "Design for safety" is every bit as important as—and is, in fact, part of—design for manufacturing, assembly, and supportability.

The Department of Labor defines three primary measures of workplace safety: (1) Recordable Incident Rate, (2) Lost Workday Case Rate, and (3) Lost Workday Rate.[5] They are not just indicators of safety in the workplace, but actionable metrics that are the responsibility of every person in the workplace. If an employee senses a potential accident, they should intervene immediately if appropriate and at the least, bring it to their manager for appropriate action.

The Boeing Company instituted a Go-4-ZERO safety program, where all leaders signed a Safety Promise, their personal accountability to proactively prevent safety incidents and injuries. This program was championed at the very top of the company and driven down. The premise was simple: all employees would go home from work in the same condition in which they came to work. I made it a point at my weekly leadership tag-up to review the safety metrics above; specifically every recordable, lost workday rate, and lost workday case rate at every site. Leadership Teams throughout the company, including mine, took appropriate action to address safety related findings.■

Chapter Eleven Technical Takeaways

- Your job: to support the business and protect the enterprise. The buck stopped with me when it came to technical-decision making.

- Engage with your team in a proactive and collaborative way to gain a perspective on the technical health of your programs; I recommend deep-dives and the "trust but verify" approach.

● Systems-level and critical thinking is essential; understand the metrics needed to evaluate program health—and affirm that the team is using them to manage on a daily, weekly or monthly basis.

Chapter Eleven Leadership Takeaways

● The chief engineer is the guardian of technical excellence for a project, program, or a portfolio of programs. Your role: to teach and reward people and teams which exemplify technical excellence.

● Create a culture of "Help Needed," "Don't Shoot the Messenger," and it's okay to "Bring Bad News Forward."

● How best to implement an Engineering System in your organization? You want to leverage people across the business and enterprise when needed and in a collaborative way.

● Safety is everyone's responsibility—but as a leader, you must take charge, make sure that people have a safe work environment.

1.*The Malcolm Baldrige National Quality Award* is a management system framework for achieving Performance Excellence. Administered by the National Institute of Standards (NIST) within the Commerce Department, the primary objective of this Presidential-level recognition is to improve the performance and competitiveness of U.S. companies, schools and non-profit organizations.

The category criteria, at the time we used it in Aerospace Support, included

Leadership, Strategy Development and Deployment, Customer Focus and Relationships, Process Based Management, Human Resources and Business Results. The criteria apply to most types of organizations (e.g., manufacturing, service, health care, education and non-profit). I felt it was a great time for the emerging business unit, as it was growing so quickly, to utilize the Baldrige criteria to help provide the organization with a foundation upon which to execute its contracts and grow.

From my perspective, the biggest benefit of using the Baldrige framework is that the applicant receives feedback on how to improve the company or organization from a team of independent examiners outside the company and industry. As a note, I applied for and was accepted to be a Baldrige Examiner in 2004. It was an excellent personal development opportunity and it helped me to become a better leader of large organizations.

2. Welch, Jack, JACK—*STRAIGHT FROM THE GUT*, (New York, NY Warner Brothers Books, 2001) p. 205

3. Gasparro, Annie, *ConAgra Targets Cost Cuts*, Wall Street Journal, p. B5, October 05, 2015

4. OSHA, Safety and Health Related Topics, *Business Case for Workplace Safety*, https://www.osha.gov/dcsp/products/topics/businesscase/ (Retrieved 15 SEP 2015)

5. A Brief Guide To Recordkeeping Requirements For Occupational Injuries And Illnesses https://www.osha.gov/archive/oshstats/guidelines.html, Retrieved 30 NOV 2015

CHAPTER TWELVE
ETHICS IN ENGINEERING

Above all, we must realize that no arsenal, or no weapon in the arsenals of the world, is so formidable as the will and moral courage of free men and women. It is a weapon our adversaries in today's world do not have.

Ronald Reagan
40th President of the United States

E thics is about personal accountability, your character, and reputation. It's a willingness to accept responsibility for your work. The consequences of unethical behavior can be damaging to you as an individual—loss of employment, monetary fines, prison time. And, damaging to a company not just with fines, but of greater significance, being disqualified to compete for future contracts.

Examples of unethical behavior include:

- Lying about your work experiences and expertise to your company or customers.

- Mis-charging your labor time to one contract when working on another.

- Improper engagement with existing or potential suppliers or customers.

- Releasing an engineering drawing or a product knowing it does not meet all technical and regulatory requirements. Your signature is your personal accountability, that the product meets all customer, regulatory and safety requirements.

- Wrongly affirming that required inspections and tests had been conducted.

- Falsifying reports or public information releases, hiding known problems, or incorrectly claiming that corrective actions had been taken.

- Companies knowingly hiding product integrity issues or concerns well in advance of a recall or notifying customers.

A recent article in the Wall Street Journal (WSJ) highlights the example of a foreign automotive manufacturer in crisis over cheating on emissions testing. The company said that as many as 11 million vehicles, world-wide, carried a "defeat device," software that reduces tailpipe emissions when the car is being tested, but not when on the road."[1] In a subsequent WSJ article, the company "blamed a damaging emissions crisis on a 'chain of mistakes' that began with the company's diesel push into the U.S. in 2005 and a 'culture of tolerance' for rule breaking that allowed the deception to continue for a decade."[2] Consequences? Beyond a seriously compromised reputation, the U.S. alone is seeking something like $48 billion in fines. You get the idea . . .

Engineers make decisions based on facts and data. As a profession, we rely on our education, background, work experiences, company processes, analytical tools, laboratories, and simulations to help us do our jobs as efficiently and effectively as possible.

Yes, we at times make assumptions but not without opportunities for follow-on testing and validation.

However, pressure to meet technical requirements, schedule, and cost commitments can sometimes drive wrong behaviors. For example, approval to release a drawing before all analyses had been completed, moving from one phase of development to another before all criteria have been met, or approval to proceed into ground or flight test before all checks have been completed and approved through a rigorous safety review board. It's these expectations and behaviors that drive the need for engineering excellence, ethics in engineering and technical decision making without compromise for product safety, integrity and quality.

I should note, there are often acceptable interim solutions. For instance, to help move programs forward in development before an aircraft is ready for full flight-envelope testing, you could impose temporary operating limits (TOL) or flight restrictions. Analysis is conducted at the TOL level. As long as a positive margin exists and is documented, the aircraft is cleared to fly within the bounds of the restriction.

As an engineering leader, I felt it was my job to create the environment where my team could bring forward any issues or concerns, which they felt were in conflict with the Code of Ethics. At our monthly strategy meetings, we often led off with "An Ethical Moment." One member of the team would have been designated to tell a story involving an ethical situation: the issue, the actions taken, and the outcome. Then, the person who made that presentation would tag someone to share a story at the next meeting. This may have involved a bit of gamesmanship, perhaps a chance to put a friend on the spot . . . but . . . the selectee would now have to do some research, prepare a presentation, practice, and literally "own" the issue. You could call this, "total immersion," a great tactic for getting team members fully involved.

Let me share two of the more frequently-cited examples of ethical violations—not just from our monthly meetings, but from any training session on the topic in almost every corporation in the nation . . . including, of course, Boeing, which has an on-going program especially for "new hires."

Number One: you open your mail (or email) and see a document sent by a government customer, which clearly was meant for a competitor and stuck in the wrong envelope (or sent to the wrong address). What should you do? Answer: Stop reading, and contact the Legal Department immediately.

Number Two: Boeing does business all over the world, working with foreign customers and suppliers. Different countries, different cultures. As a result the U.S. Government put in place the Foreign Corrupt Practices Act (FCPA) to ensure against unlawful or unethical acts that could influence a business outcome.[3] Such as, what is a gift and what is a bribe? Just assume, there is no such thing as a "gift" to a foreign customer. It is important for anyone working internationally to fully understand the FCPA.

During your career, you will be faced with hard choices, some of which may have significant and lasting impact on you—and on the people with whom you work. Ensure that those choices reflect the qualities for which you want to be known. The decisions you make in life, matter. ■

Chapter Twelve Takeaways

- Understand how to handle procurement integrity situations and other laws associated with doing business, such as the Foreign Corrupt Practices Act, both in the United States and international. Make sure people are trained on a periodic basis.

- An engineer's signature on a drawing is their personal accountability and stamp of approval.

- The work you do and the decisions you make matter.

- Ethics and personal accountability in engineering mean there is no compromise for product safety, technical integrity and quality.

- Look for teachable moments and share with members of your team.

- The importance of having an open culture in the work environment, where people are encouraged to bring issues forward without fear of retribution, is a significant enabler toward preventing ethical violations.

---

1. Boston, William, VW, Wall Street Journal, p. A10, 05 October 2015

2. Boston, William, Varnholt, Hendrik, Sloat, Sarah, VW Says *'Culture' Flaw Led to Crisis,* Wall Street Journal, p. B1, 12 December 2015

3. Gunn, Thomas   *Gunnsights*, pp. 16-17, (Annapolis, MD: Naval Institute Press, 2007)

CHAPTER THIRTEEN
CAREER MAPPING

What is the recipe for successful achieve-
ment? To my mind there are just four essential
ingredients: Choose a career you love, give it
the best there is in you, seize your opportuni-
ties, and be a member of the team.

Benjamin Franklin Fairless
American Businessman

Throughout my career, I have watched people take different approaches to managing their own careers. Some developed detailed plans, what job level/responsibilities they expected laid out against a specific timeline, augmented by outside experiences such as graduate school or other continuing education. They knew where they were going and how they planned to get there. At the other extreme were those who did not "plan" but assumed that always demonstrating a high level of job performance combined with a positive, enthusiastic, and willing and able-to-work together attitude would do the talking for them.

I fell into the latter camp. My mindset, based on my upbringing and the work environment, was to work hard and assume that success would follow. There were times such as my rotation into Structural Dynamics and Loads in 1983, the *Harrier* Wing Team lead in 1989, or Information Systems in 1996; where I was tapped to do these assignments and frankly never saw any of them coming. Other times, my managers asked for volunteers to leave existing assignments to consider opportunities on emerging programs. Only once through my career did I have to interview for a job.

Today's generation and workforce as a whole, based on my experience, is much more assertive and ambitious, which I think is great. Today's leaders are much more approachable and accessible than they were when I was an early career, even mid-career engineer. I spent a lot of time as an executive, mentoring people individually and talking with groups such as emerging leaders in many different venues. Developing people and helping them achieve their goals and aspirations were the best part about being in a leadership position.

Career Mapping Recommendations and Considerations

1. Develop a career plan that is in the middle of the two extremes. If you are an early career employee (up to eight years or so), consider approaching your leadership and mentors and express interest in career broadening job experiences. For example, if you are working in the field of structural analysis, you might suggest moving into an area of special focus, such as metallic design (static load stress analysis), durability and damage tolerance (fatigue or crack initiation, and crack growth or fracture mechanics), composite materials design, or Finite Element Analysis. Further, there are different areas of structural analysis: wing and empennage structure vs. fuselage structure vs. systems (e.g., fuels, hydraulics and electrical). It's important to know the nuances of each, and especially the differences in failure modes, to ensure that you will be making the right analytical checks.

Another example of an early career development is becoming a technical lead, where one can advise, assist and check work of more junior engineers. Adjacent functions include structural design, structural dynamics and loads, and liaison engineering or factory support. These kinds of experiences prepare you not only for a bigger and better assignment in a technical role, but in a leadership one as well. A strong technical foundation is critical to both your job performance and advancement opportunities.

To further highlight this point on engineers having a strong technical foundation, let me share an example of what was to be one of my last challenges prior to retirement. One day, a VP/GM called and asked me for help. During a ground test, measured loads on an external pylon were significantly higher than was predicted in one loading direction. As discussed in Chapter Four and then reiterated in Seven, new aircraft structure is tested to the critical design load conditions. Because this finding during a ground test exceeded what was tested, the customer wanted to test these higher loads on a full scale static airplane to make sure that the aircraft's internal back-up structure could handle them. The aircraft test article was not readily available and it would have taken a lot of time and money to get it ready for such a test. Analytical techniques would not be enough to convince them to validate the approach, and I understood their position based on prior assignments, so their concern about methods of verification was well understood. However, given the localized area of the implication of higher loads, there was an alternative that might be the right decision for all stakeholders.

To make a long story short, I recommended the addition of a structural beam to handle the increased load, consistent with the load direction. This enabled the aircraft interface previously tested to not see an increase in loads over what was previously tested. Because the addition of the structural member was statically determinate (this means there was no ambiguity on aircraft-pylon interface loads), sizing this additional structural member was very straightforward, thereby making it acceptable to verify the design for the increased load by analysis. The customer liked the approach and agreed that a full-scale test, which would have had significant cost implications, was not necessary. I attributed this outcome as a result of having built a strong technical foundation in structural analysis and an understanding of the customer throughout my career.

As a mid-career employee, you may be interested in obtaining experience in a different function outside engineering, such as supplier management, manufacturing and quality, program management, or business development. Assignments in other functions provide employees with a fresh perspective for dealing with work products from engineering.

In addition, ask what other work experiences and educational opportunities exist for those desiring careers in program management and executive leadership. Be flexible with timelines. Sometimes there may be no job openings, so be patient. This kind of conversation comes across as more genuine than "I want to be a manager in five years, director in ten years, etc." Your leadership knows you want promotions and opportunities to advance.

Chapter Eight describes the importance of "supplier technical oversight" throughout the product development life cycle. Some of the Case Studies in Chapter Ten highlight the need for more effective insight and oversight of the supply chain. A great rotational assignment for engineers looking for leadership opportunities, while utilizing and building technical skills, is a role in your company's supplier or supply chain management organization. With companies relying more and more on their supply chain, this remains one of the best rotational or even career changing assignments.

2. Consider work assignments in different business units or programs within different life cycle phases. For example, in today's DoD time horizon, contractors are expected to utilize a technology demonstrator aircraft to demonstrate some of the most important requirements as part of a future selection process. Processes, such as prototyping, and team structures may be different for this phase of the product life cycle.

3. Make good first impressions with people. As I said in a previous chapter, you never know who will weigh in on your next potential opportunity, including people who may or may not be in your chain of command. During a structured interview, the interview panel will frequently include a stakeholder associated with a work group. Also, you might be surprised to know who might be weighing in on your merit review.

4. Remember NFL Coach Vince Lombardi's admonition, "Luck is what happens when preparation meets opportunity." So, prepare for interviews, whether formal or informal. In Chapter Six, some typical questions and general thoughts are provided to help you have the best interview you can. Ask peers to perform a mock interview so you can be concise with your answers. Be prepared to give examples of your work process and success, including details—perhaps, what was the key to your success. Preparation, though, also includes an ability to demonstrate your understanding of the job you seek, and knowledge of the program—including challenges being faced.

Many times, a panel member might not know some of the people being interviewed. Call it, the ultimate opportunity to make a good first impression. A couple of times through my career (and I was on countless interview panels), candidates would be unprepared and they weren't afraid to say so. For the couple of times this happened, the candidate did not get the job, not because they had a cavalier attitude but because they didn't address the questions very well. Further, when these individuals would come by for a career discussion, let's just say that their interview left a lasting impression on me. I hope that I have been able to convince readers that "preparation" is key to almost anything.

5. Demonstrate a track record of a high level of job performance; including, how well you work together with others. Your ability to deliver results and how you get those results is directly within your control. Whether you pursue a technical or management career path, you need to have good interpersonal skills, such as the ability to communicate effectively, be a good team player, be able to lead people, and, especially, to handle "change." You certainly don't have to have a management title to be an effective leader. Chapter Six provides additional information hiring managers consider when making selection decisions for new assignments.

6. Develop effective mentoring relationships. When thinking, by whom you would like to be mentored? Consider:

- Feedback from your peers about who they consider to be an effective mentor.

- Someone who has inspired you directly or indirectly.

- Mentors within and outside engineering, so you get the benefit of learning what's important across the business.

As a mentor, myself, I would ask the mentee to send me a note with two or three topics they wanted to discuss or questions they wanted to ask at our next meeting. This gave me time to really think through my response, to give my mentees the best feedback possible. Table 13.1 (next page) provides a Career Path Framework. ∎

Chapter Thirteen Takeaways

- YOU have the responsibility to take charge of your own career.

- Plan to take assignments that are out of your comfort zone to broaden your knowledge of the business.

- Identify and secure mentors to help you with career guidance and suggestions.

- Establish a positive track record of performance and of your ability to work with people—and, as a member of a team.

Table 13.1 / Career Mapping Framework

Career Phase	Objectives	Action Plan (Example)
Early Career (Up through 8 years) Plan for 2-4 assignments- 2 to 3 years each	• Build a Strong Technical Foundation • Broaden Technical Knowledge Through Assignments Within Engineering	• Advanced Engineering Degree • Continued learning programs • Technical Lead Engineer • Consider moving to programs in different life cycle phases • Mentoring relationships
Mid-Career (8 through 15 years) Plan for 2 to 4 assignments- 2 to 3 years each	• Understand Product Life Cycle and Value Stream of Products and Services • Pursue Career Broadening Assignments Within or Outside of Engineering • Hone in on Career Paths	• Start developing customer relationships • Continued Learning Opportunities • Expand Technical Capabilities • Seek Opportunities in Different Business Units, and Team Structures • Strengthen Leadership Skills, Pursue MBA Pending Career Interest • Build Relationships • Mentoring Relationships
Mid-Career+ (15 -25 Years) Plan for 3 to 4 assignments- 2 to 3 years each	Typical Options Include: Engineering —Technical Path Engineering —Leadership Path Supplier Management Program Management	• Continue to Build Customer Relationship and Understand Wants and Needs • Continue to Seek Assignments that Broaden Leadership and Technical Capabilities and Skills • Always Be Prepared to Step-up to new opportunities
Post Mid-Career+	Hone in on Skills/ Interests	• (Same as Mid-Career+)

CHAPTER FOURTEEN
EXECUTIVE ENGINEERING LEADERSHIP

I always "slept better" because of a great high quality chief engineer. Designing, building, testing, operating and sustaining a complex weapons system is always a balance of risk. With a high caliber, technically savvy, customer focused, people developing chief engineer, she/he would always deal with the many challenges, issues, misunderstandings, and curve balls every day throws at us.

Senior Boeing Business Leader

I have been fortunate to have had so many diverse assignments, each broadening my technical and leadership skills, as I acquired knowledge and understanding of the aerospace business. Valuing and developing people, having a focus on the customer, embracing strong core values, recognizing the importance of first time quality, leading teams on a journey of technical excellence, building effective relationships, the ability to work together with others, and being prepared every day were instrumental to delivering results in every job I had. Whether it was an engineering, manufacturing, quality, information systems or program management role….working on a fixed wing, rotorcraft or commercial derivative aircraft…in every phase of the Product Life Cycle, for a domestic or international customer, it didn't matter.

While I enjoyed every job I have had, it was clear to me that being a chief engineer felt right. Engineering was my strength, my passion and where I could provide the most value for the company . . . and, where I had the opportunity and privilege to work with people associated with every military aircraft in Boeing's Defense business portfolio.

In whatever job you're in, I encourage every person to know the "key success factors." For a business unit chief engineer, they were as follows:

It's About Leadership

A chief engineer's job is about creating and communicating a vision across multiple time horizons, developing strategy, obtaining resources, and rallying a team to work together to achieve that vision. It's about selecting and motivating people at all career stages, setting expectations and inspiring them to perform. It's about making sure you are making the time necessary to help them reach their potential through the right developmental opportunities and job assignments. It's about building positive relationships with people who can help accomplish team, program or business unit objectives. It's about delivering on customer commitments in a profitable way that meets shareholder requirements. It's about accomplishing the work that needs to be done when it needs to be done. It's about creating a productive work environment where people are encouraged to innovate, whether on a new widget or new ways of doing business. It's about creating frameworks so others can achieve desired results. It's about recognizing that people are a company's biggest asset, given the human or intellectual capital in the workforce. It's about putting the right people in the right job at the right time. It's the leader's job to know their people and leverage their strengths to the fullest. It's about setting priorities and effectively handling conflict resolution. It's about relating to people at all levels within the organization. It's about being approachable and wanting to hear from people on their ideas to improve first time quality or accelerate the journey of technical excellence.

It's about proactive succession planning. Programs can go on for fifty years or more, but people come and go. It's a leader's job to build a strong pipeline of people who can maintain continuity, running a healthy program, satisfying both customers and shareholders.

It's About Technical Excellence

A chief engineer's job is about supporting the business by designing, building, testing and fielding products that enable our customers to accomplish their missions. It's also about protecting the enterprise. It starts with a culture of ethics in engineering and personal accountability. It's about making technical decisions based on facts and data. It's about managing technical risk and bringing opportunities to fruition. It's about creating a work environment that (1) establishes a culture of asking for help when needed, affirming that "we won't shoot the messenger" and we actually welcome bad news, and (2) that enables empowered and adaptable teams, but within an engineering system that provides "checks and balances" through collaborative, "trust but verify" engagements. It's about having situational awareness of program health and knowing when to dive deep and take appropriate action to help recover adverse performance trends. When things go wrong, ask, what processes broke down? That approach puts people more at ease in difficult situations and helps get to the root of a problem. It's about looking at process health holistically (across teams and programs) that contribute to poor quality and taking appropriate actions to fix and or strengthen them. It's about not closing a gated milestone, nor allowing a ground or flight test to proceed before its ready.

It's Also About Providing Proven Processes and Tools

Processes: How people design, build and test. Processes and process discipline are critical to success of a business unit, program or project. Processes are not intended to be bureaucratic nor stifle innovation. To the contrary, systems level thinking around processes is about applying higher level methodologies and approaches to achieve first time quality designs. Process development and deployment is about combining systems level thinking with the "how-to" steps. This includes defining suppliers, customers, inputs and outputs of the process, and providing adequate checks and balances. Last, it's about measuring process health. When things get measured, performance generally will improve.

Tools are automated programs that analysts use to ensure the design meets the requirements, whether calculating structural margins of safety, gain and phase margins in control law development, or power and cooling requirements for mission avionics. First time quality in engineering depends on ensuring tools are mature and production ready, that they are aligned with processes, and that analysts have an appropriate level of training. As an engineer, it's your job to flow requirements to the team responsible for tool development and implementation.

It's Also About Promoting Science, Technology, Engineering and Mathematics (STEM) For Future Engineers

I am a big proponent of promoting STEM education to students, especially those with high aptitudes in mathematics and science. I want to expand the pipeline of students pursuing continuing or college education, and encourage them to consider careers in engineering, the sciences, technology and related fields. Different sources of data say there will be a nationwide shortfall, possibly as high as one-million people in STEM related fields entering the workforce by 2020.[1] Further, the data shows lots of untapped potential, that is, there are students that have the academic potential, even people graduating with STEM degrees but who for whatever reason are not interested in pursuing careers in education, technology, and the sciences.

What can we collectively do as engineers? We can volunteer our time in elementary classrooms once or twice a month, sharing stories of what got us interested in our careers, showing the students videos of our company's products, or just telling them what we do. We can help with classroom work assignments, science fair projects, or First Robotics competitions. Every engineer has the opportunity to make a difference, by helping students better understand careers in STEM-related fields, highlighting the importance of math and science. A team from my company visited 4th, 5th, and 6th grades at a local school and ignited student interest with cool videos of aircraft flying, and let the students listen to a test pilot talk about flying jet aircraft. The students we engaged had so many questions. It was a very energizing hour, for the students and the engineers.■

Chapter Fourteen Key Takeaways

- A chief engineer has the responsibility to lead people, to ensure a culture of technical excellence, to communicate with clarity, and to put in place a work environment that values ethics, integrity, and quality.

- Successful chief engineers at the end of the day must have great judgment due to the depth and breadth of situations you handle, and decisions you make, everyday.

- Engineers must promote science, technology, engineering, and mathematics to students, planting, if you will, seeds for the future.

---

1. Charette, Robert N., The STEM Crisis is a Myth: An Ongoing Discussion, Posted 30 August 2013 http://spectrum.ieee.org/static/the-stem-crisis-is-a-myth-an-ongoing-discussion (Retrieved 16 Nov 2015)

CHAPTER FIFTEEN
CHOOSING TO BE UNSTOPPABLE

Physical fitness is not only one of the most important keys to a healthy body, it is the basis of dynamic and creative intellectual activity.

John F. Kennedy
35th President of the United States

M y job was physically and emotionally challenging. Travel was excessive and over the last ten years of my career, I was doing work all around the world. The days were long and it was often hard to remain focused, especially by the end of a hectic day. The breadth of my responsibility was huge, wrapped around the many programs and contracts being executed, the size of the engineering workforce, and the number of sites where work was being performed. There were always commitments to meet, meetings to attend, challenges to be dealt with, and people to answer (or, answer to); as noted, 80 to 100 emails per day. While I loved the job, the pace was grueling and work-life balance was, well, not in balance.

Some people counter job stress by taking a "time out" (or. . ."time off") or find their own path to recharge the batteries. For me, the priority was staying healthy; otherwise, I would not be of much good for my family, my job, my people, and my teams. I approached "health and wellness" much as if I was on a championship sports team preparing for the next competition. Being in the best physical shape had a direct and positive correlation with my attitude, teamwork, and job performance.

I have always been "physical," playing tennis and lifting weights. But when I took on a personal trainer, whose mantra was "Choose to be Unstoppable," my regimen became intense and focused. Strength, conditioning, and "strongman" training kept me energized through the day and gave me the edge to be as productive as possible. In addition to conventional barbells, dumbbells, and kettlebells, we flip tires, and lift, squat, and carry sandbags and beer kegs, pull and push sleds loaded up with weight, climb ropes, and more. There are lots of pushups, pull-ups, sit-ups, burpees, sprints and other speed drills. This level of physical activity and challenge was a big driver to my level of energy, focus and productivity.

I worked with many people who were just as passionate about their sport. One was an Ironman competitor, another a mountain climber, some people were into bodybuilding, martial arts, and short and long distance running. We would enjoy talking about and learning about each other's sports, and share experiences and training methods. My trainer emphasizes, there is a positive correlation between success—in work, and in life—and a regimented fitness program. I certainly agree with articles I've read, which advise that exercise releases endorphins that provide a calming effect and boost confidence and energy, all day long.

The best "physical fitness" advice, from both my trainer and doctor, was to find ways to keep moving. Sitting at desks, on

Figure 15.1 Donovan's Strength and Conditioning with Head Strength Coach Tommy Donovan (pictured on left)

airplanes and in meetings for a significant part of the workday, day after day, week after week, and year after year, will over time limit your mobility. Do the best you can to walk as much as possible, take stairs instead of elevators, get up from your desk and move around the office.

How do you start . . . or end . . . your day? I do suggest making time for yourself, whatever the activity you find that energizes you and gives you an outlet for daily stress . . . and keeps you healthy. ■

Chapter Fifteen Takeaways

• Choose to be unstoppable in a way that works for you. It will help keep you healthy.

• Vigorous physical exercise continues to be a great way that energizes me for the day.

• As leaders, we have tough jobs, some people call it a combat sport. There are many ways people rejuvenate and recharge their batteries. From first-hand experience, physical exercise is a great way to beat down the stress of the job.

• Spend time engaging with others and learn about their activities—you might find something you might be interested in trying. It's also a great way to strengthen relationships and build teams.

CHAPTER SIXTEEN
CALL TO ACTION

Action is the foundational key to all success.

Pablo Picasso
Artist

Many leaders across the company would conclude classes, held in different venues, with a "Call to Action." The purpose: to circle back with the classes and ask questions that would get them thinking about how they might apply what they learned, whether from me or through a dialogue with others in the class. As I come to the end of this text, let me leave you with some things to think about—regarding your career, or your life.

Call to Action: Considerations/Questions for business and engineering Leaders.

- If you are a business leader for an engineering company, what expectations are you setting for your engineering leadership?

- What are you doing or have you done to create a culture of first-time quality and technical excellence?

- How have you emphasized the importance of ethics and personal accountability in your organization?

● Have you ever asked your direct reports to what extent they are inspired by you?

● What processes have you developed and deployed in your business unit or team to improve productivity and quality?

● How are you sharing best practices and lessons learned in your business?

● How much time are you spending on people, executing today's business vs. positioning the business for the future?

● How do you translate strategy-to-action plans that include people, processes, tools and technology?

● Are you considering manufacturing, ergonomics, and supportability in the design of your products.

● What are you or your company doing to promote STEM to students who have the interest and aptitude in mathematics and science?

Call to Action: Considerations/Questions for graduating students ready to enter industry, early and mid-career engineers, and students aspiring to be engineers.

● How are you building your technical foundation?

● What is your continuing education or learning plan?

● Are you considering manufacturing, supportability, and ergonomics in the design process?

● Familiarize yourself and learn from examples of unethical behavior in engineering. I read about them too often in the media.

- Don't hesitate to ask peers, technical leads, or managers about any question or concerns.

- What are you doing to promote STEM to students who have the interest and aptitude in mathematics and science? ∎

EPILOGUE

D ear reader, as you are progressing through your career, continue to ask yourself, "What legacy do you want to leave?"

As for me, I had the career-long opportunity to engage with people, teams, and emerging leaders throughout. I developed and mentored many of them, including people within and outside of engineering, influenced careers with new and different assignments, taught "Leaders Teaching Leaders" classes at the Boeing Leadership Center, and simply led by example. When I see many of the leaders throughout the defense-side of The Boeing Company, today, I feel like I did leave a legacy— through people. Throughout my career, I was a guest lecturer at multiple universities, including Purdue University, Iowa State University, University of Oklahoma, Washington University and Southern Illinois University, teaching students on the A-10 wing (referenced in this book), along with "What Success Looks Like in Industry," and "Ethics in Engineering." I enjoy engaging with college students; not sure if I left any kind of legacy with them, but hope that I encouraged some to make the step from college and take their education to work in the aerospace industry.

However, my legacy also continues through decisions, which I made about aircraft, now flying worldwide and which will be flying safely and reliably for decades to come. Often, I look up into the sky . . . and see some of my mechanical progeny, at work.

A new chapter in my life started on October 12, 2013 and since, may I say? Spending time with relatives, friends, and my physiological progeny—children and grandchildren—has been very fulfilling; I sense there may be one or even two future engineers in about fifteen years or so. It's been fun having the time to go to their extracurricular activities.

On a personal note, I am in the best physical shape of my life. Exercise is one of my favorite things to do. Physical fitness for me is a life-long commitment.

Staying engaged with the aerospace engineering profession is high on my priority list. I have had the opportunity to consult for Boeing and one other company since I have retired and plan to continue doing so as the opportunities present themselves. Writing this book has given me the time to reflect on my work and life experiences, and share thoughts that might be beneficial to others. I have had opportunities to volunteer my time by providing civic and community service. Teaching, and more importantly, interacting with others in various venues, whether engaging with teams, emerging leaders, or in a classroom, has always been a passion. I am currently an Adjunct Faculty Member at Lindenwood University in the School of Science / Department of Mathematics. My objective: to teach and prepare students, the next generation of the workforce, for success in industry or their chosen field of interest.

I am so grateful for my career at Boeing. I was blessed to have such a range of experiences, and the opportunity to make a difference. The people I worked with and friendships that I made were the best part. Given all of that, I wouldn't trade my retirement decision for anything … the balance in my life today is priceless.■

BIBLIOGRAPHY

"A Brief Guide To Recordkeeping Requirements For Occupational Injuries And Illnesses" https://www.osha.gov/archive/oshstats/guidelines.html, (Retrieved 30 NOV 2015)

"Apollo 13" NASA, 8 July 2009 https://www.nasa.gov/mission_pages/apollo/missions/apollo13.html#.Vkyb_N5CjjA (Retrieved 18 NOV 2015)

"Base Realignment and Closure (BRAC)" https://en.wikipedia.org/wiki/Base_Realignment_and_Closure (Retrieved 21 SEPT 2015)

"Boeing 737 AEW&C *Wedgetail* Early Warning Aircraft, Australia," http://www.airforce-technology.com/projects/737aewc/ (Retrieved 22 NOV 2015)

"Boeing, *Chinook* and *Chinook* Through Life Customer Support" http://www.boeing.co.uk/products-services/boeing-defence-uk/aircraft-services-support/chinook-tlcs.page? (Retrieved 12 NOV 2015)

Boeing Frontiers Article, OCT. 2009 [Lean+] "Simple As…" http://www.boeing.com/news/frontiers/archive/2009/october/i_ids01.pdf (Retrieved 12 NOV 2015)

Boston, William, "VW," Wall Street Journal, 05 OCT 2015

Boston, William, Varnholt, Hendrik, Sloat, Sarah, "VW Says 'Culture' Flaw Led to Crisis," Wall Street Journal, p. B1, 12 December 2015

BusinessDictionary.com. http://www.businessdictionary.com/definition/ergonomics.html#ixzz3z1KiatOc (Retrieved 02 February 2016)

Butler, Amy, "Defense Development Timelines Thwarted by Poor Software Planning," Aviation Week's Daily and Defense Report, 10 JUL 2015

Charette, Robert N., "The STEM Crisis is a Myth: An Ongoing Discussion," Posted 30 August 2013 http://spectrum.ieee.org/static/the-stem-crisis-is-a-myth-an-ongoing-discussion (Retrieved 16 NOV 2015)

Columbus McKinnon Corporation, "When Luck Meets Opportunity (Vince Lombardi)," 17 JAN 2011 http://blog.cmworks.com/when-preparation-meets-opportunity/ (Retrieved 21 SEP 2015)

Deckelbaum, Isadore http://genealogy.caroldeckelbaum.com/memoir-izzy1.html, http://genealogy.caroldeckelbaum.com/memoir-izzy2.html, 1976-1977 (Retrieved 16 NOV 2015)

Department of Defense—Systems Management College Systems *Engineering Fundamentals*. Ft. Belvoir, VA: Defense Acquisition University Press, January 2001 http://www.dau.mil/publications/publicationsdocs/sefguide%2001-01.pdf (Figure 7.1, p.65)
(Retrieved 31 January 2016).

Dorr, Robert F, Defense Media Network, "Gulf War 20th: F-15 *Eagles* Were the Deadliest Birds of Desert Storm," 07 JAN 2011 http://www.defensemedianetwork.com/stories/f-15-eagles-were-the-deadliest-birds-of-desert-storm/ (Retrieved 05 DEC 2015)

"Elite UK Forces." http://www.eliteukforces.info/air-support/7-Squadron/chinook.php Retrieved 12 NOV 2015)

Fabey, Michael, "Ford Based On Unrealistic Business Case, GAO Says." Aviation Week's Daily and Defense Report, 05 OCT 2015

Gasparro, Annie, "ConAgra," Wall Street Journal, 05 OCT 2015

Gerald R Ford Class (CVN 78/79)—"US Navy CVN 21 Future Carrier Programme, United States of America," http://www.naval-technology.com/projects/cvn-21/ (Retrieved 12 NOV 2015)

Gunn, Thomas, *Gunnsights*, Annapolis, MD: Naval Institute Press, 2007

Joint Publication 1-02, Department of Defense *Dictionary of Military and Associated Terms,* 8 NOV 2010 (As Amended Through 15 OCT 2015) http://www.dtic.mil/doctrine/new_pubs/jp1_02.pdf (Retrieved on 15 NOV 2015)

Lavretsky, Eugene and Wise, Kevin, *Robust and Adaptive Control With Aerospace Applications,* Chapter 1, 2013

Martin, James R. Ph.D., "CMA Learning and Experience Curves," http://maaw.info/LearningCurvesMain.htm Retrieved on 12 NOV 2015

Notter, Jamie, "Generational Diversity in the Workplace," http://www.multiculturaladvantage.com/recruit/group/mature/generational-diversity-in-workplace.asp (Retrieved 05 DEC 2015)

OSHA, Safety and Health Related Topics, "Business Case for Workplace Safety," https://www.osha.gov/dcsp/products/topics/businesscase/ (Retrieved 15 SEP 2015)

Wallace, Kelly, "A year of paid parental leave: Vital but how likely?" CNN article, 10 AUG 2015 http://www.cnn.com/2015/08/10/health/paid-parental-leave/ Retrieved 10 Aug 2015)

Welch, Jack, *JACK—Straight From The Gut*, New York, NY, Warner Brothers Books, 2001

GLOSSARY

AETC	Air Education and Training Command
AEW&C	Airborne Early Warning & Control
ALC	Air Logistics Centers
BAE	British Aerospace Systems
BDS	Boeing Defense, Space and Security
BMA	Boeing Military Aircraft
BRAC	Base Realignment and Closure
CDR	Critical Design Review
CONOPS	Concept of Operations
COTS	Commercial Off-The-Shelf
CPAR	Contractor Performance Assessment Report
CTF	Combined Test Force
DAU	Defense Acquisition University
DCMA	Defense Contract Management Agency
DFMA	Design for Manufacturing and Assembly

EMALS   Electromagnetic Aircraft Launch System

EMD   Engineering, Manufacturing and Development

EVM   Earned Value Management

FCPA   Foreign Corrupt Practices Act

FTQ   First Time Quality

GAO   Government Accountability Office

GS&S   Global Services and Support

I&CO   Installation and Checkout

ICD   Interface Control Drawing

IM   Instant Messenger

IOT&E   Initial Operational Test & Evaluation

IPD   Integrated Product Definition

IPT   Integrated Product Teams

ISR   Intelligence, Surveillance, and Reconnaissance

KPP   Key Performance Parameters

LRIP   Low Rate Initial Production

MDA   McDonnell Douglas Aerospace

MTBF	Mean Time Between Failure
NAVAIR	Naval Air Systems Command
NDI	Non Development Items
OEM	Original Equipment Manufacturer
OPEVAL	Operational Evaluation
PACS	Programmable Armament Control System
PDM	Programmed Depot Maintenance
PDR	Preliminary Design Review
PM	Program Manager
PRR	Production Readiness Review
QPD	Quality Processes Division
RAA	Responsibility, Authority and Accountability
RAAF	Royal Australian Air Force
RCCA	Root Cause and Corrective Action
RFP	Request for Proposal
SCD	Source Control Drawing
SPO/SPD	System Program Office/System Program Director

SRR	Systems Requirement Review
STO	Supplier Technical Oversight
TIM	Technical Interchange Meeting
TLCS	Through Life Customer Support
TOL	Temporary Operating Limits
TRR	Test Readiness Review
UARRSI	Universal Aerial Refueling Receptacle Slipway Installation
VP/GM	Vice President/ General Manager
V/STOL	Vertical/ Short Takeoff and Landing
WARP	Wing Aerial Refueling Pod
WBS	Work Breakdown Structure

Glossary of Technical Terms

Economic Profit—defined as Net Operating Profit after Taxes (NOPAT) minus the cost of capital associated with carrying its net assets. Therefore, with Services, net assets, such as PP&E are lower in a business that has relatively large engineering and manufacturing related assets and thus enabling a higher level of economic profit.

First Time Quality (FTQ)—Completing the job right the first time, and without traveling technical risk downstream to test or production.

Integrated Product Teams (IPT)—These teams have full responsibility, authority and accountability (RAA) for their product, such as, a wing or a fuselage. This IPT concept encourages teamwork, real time decision-making and enables adapting to whatever change or adjustments in the plan are needed as product development evolves.

Leading Indicators—Metrics that define the health of a program and specifically highlight if planned milestones will meet the technical requirements, be completed on time and within budget. Examples include measures of:

- Critical path milestones are on track to meet scheduled date; example External Loads from Loads Team

- The number of drawings are being completed in sequence and on-time

- Percentage of Drawings that are released without requiring second effort

- Technical Supplier Specifications, documented in Source Control Drawings are released with all information the first time.

- Software Lines of Code being written to plan

- First Time Quality of software testing

Root Cause and Corrective Action (RCCA)—A systematic process for ensuring a comprehensive understanding of the root causes of a problem in order to know the fix will be effective.

Traveled Risk—Engineering work that is not completed or not completed right during the planned phase of development.

Technical Excellence—Is a journey that combines leadership, culture, strategy, organizational constructs and process, together with engineering domain knowledge and expertise, to support military aircraft programs, and protect the company from anything that conflicts with core values and principles.

Technical Risks—The exposure to a program when knowledge of technical information is unknown, or not to a level of certainty to certify a design for airworthiness. Risk is characterized by both likelihood and consequence if the risk materializes. The level of risk associated with a specific situation (see examples below) is quantified on a scale of 1 through 5, where a "1" is considered low and five "5" is high. There is a comprehensive process to manage and mitigate technical risk. Examples include:

- Use of a new material system, such as composite materials, a metallic alloy or assembly adhesively bonded together without having material allowables for structural analysis. The specific technical risk is proceeding into detailed design without having fully characterized the material for all applicable failure modes at environmental conditions defined by the technical and regulatory requirements.

- Proceeding into detailed software design without definition of the mission system requirements decomposed and flowed down to Integrated Product Teams

- New Technology planned for incorporation that is not at the desired level of maturity when the program is entering the design phase.

Technical Issues—A Technical Risk that materializes by definition becomes an Issue, which likely impacts a program's ability to meet the technical requirements, within schedule and cost commitments.

INDEX

About the Author

J eff Deckelbaum is an Aerospace Engineering Executive, who retired from The Boeing Company in 2013, working nearly 37 years in Industry. He developed a passion for airplanes as a young boy, and with strengths in Mathematics and Science, he followed that passion into the Aerospace Industry.

Deckelbaum began his career as a structural engineer on the AV-8B Harrier and advanced into positions of leadership of increasing responsibility, working on fixed wing aircraft, rotor-craft and commercial derivative aircraft. As a program and business unit chief engineer, he assembled and led people and teams that executed billions of dollars of complex, aircraft weapon systems programs for the United States Military and Allied Forces.

In addition to Engineering, Deckelbaum held leadership positions in Manufacturing or Operations, Quality, Program Management and Information Systems. He is now an adjunct faculty member at Lindenwood University.

Deckelbaum and his wife Kathy make their home in St. Charles, Missouri.